**Natural
Environment
Research
Council**

Institute of Terrestrial Ecology

Cumbrian woodlands – past, present and future

ITE symposium no. 25
Grange-over-Sands

Edited by J K Adamson

London : Her Majesty's Stationery Office

COVER ILLUSTRATION (J K Adamson)

Contrasting woodlands within Cumbria.

Top picture: Coniferous plantations are a conspicuous

feature of the north of the county where they contribute to

employment, particularly at clearfelling

Bottom picture: Deciduous woodlands are dominant in southern Cumbria

where they are widely used for recreation, including orienteering

Both pictures are inset in the outline of the

county of Cumbria

The INSTITUTE OF TERRESTRIAL ECOLOGY is one of 15 component and grant-aided research organizations within the Natural Environment Research Council. The Institute is part of the Terrestrial and Freshwater Sciences Directorate, and was established in 1973 by the merger of the research stations of the Nature Conservancy with the Institute of Tree Biology. It has been at the forefront of ecological research ever since. The six research stations of the Institute provide a ready access to sites and to environmental and ecological problems in any part of Britain. In addition to the broad environmental knowledge and experience expected of the modern ecologist, each station has a range of special expertise and facilities. Thus, the Institute is able to provide unparallelled opportunities for long-term, multidisciplinary studies of complex environmental and ecological problems.

ITE undertakes specialist ecological research on subjects ranging from micro-organisms to trees and mammals, from coastal habitats to uplands, from derelict land to air pollution. Understanding the ecology of different species and of natural and man-made communities plays an increasingly important role in areas such as improving productivity in forestry; rehabilitating disturbed sites; monitoring the effects of pollution; managing and conserving wildlife; and controlling pests.

The Institute's research is financed by the UK Government through the science budget, and by private and public sector customers who commission or sponsor specific research programmes. ITE's expertise is also widely used by international organizations in overseas collaborative projects.

The results of ITE research are available to those responsible for the protection, management and wise use of our natural resources, being published in a wide range of scientific journals, and in an ITE series of publications. The Annual Report contains more general information.

ACKNOWLEDGEMENTS

Thanks must go to J M Sykes, J C Voysey (Forestry Commission) and R G H Bunce for advice on arranging the symposium and on editing the papers, and to S Beattie, J Beckett, C B Benefield, C Benson and P A Ward for their specialist help with this volume.

J K Adamson

Institute of Terrestrial Ecology

Merlewood Research Station

GRANGE-OVER-SANDS

Cumbria

LA11 6JU

Tel: Grange-over-Sands (05395) 32264

Contents

SECTION 10

SECTION 11

SECTION 12

Preface

J K Adamson

Cumbria is a large, ecologically diverse and, therefore, scenically attractive county, and woodlands contribute in no small way to this attractiveness. However, woodlands are much more than scenically attractive. They may provide local employment, shelter for farm stock, materials for farm maintenance, fuel for the domestic hearth, and ideal venues for recreation. They have their own distinctive flora and fauna, and therefore may be highly regarded as conservation resources. If plantations of exotic tree species are included, then benefits from woodlands include employment in the transport and industrial processing of wood.

The Institute of Terrestrial Ecology's Merlewood Research Station invited speakers to present papers at a meeting in Grange-over-Sands. Those papers, which covered aspects of the past, present and future of woodlands in Cumbria, are published in this volume.

In recent years a number of organizations have produced booklets taking partisan attitudes to woodlands, particularly coniferous plantations (eg Nature Conservancy Council, Timber Growers United Kingdom). Our aim was to approach woodlands from as many viewpoints as possible, which necessitated considering both broadleaved and coniferous woodlands, but to restrict the geographical frame of reference to Cumbria. While some papers deal specifically with Cumbria, others cover topics which have a wider application but which are nevertheless highly relevant to the county.

In 1950, in *The selection of tree species*, Anderson wrote: 'Sound choice of species, and therefore sound afforestation, depends fundamentally upon the correct reading of and appreciation of all the important locality factors'. There is, perhaps, nowhere in Britain where the term 'important locality factors' should be more widely interpreted than in Cumbria; the county's diversity has led to conflicting attitudes towards some aspects of woodland management which, it is hoped, can be resolved as a result of such open discussion as was evident during the meeting at which these papers were presented.

The history of woodlands in Cumbria

J E Satchell
Lyth, Kendal

2.1 Introduction

From the New Stone Age to the recent past, woodland history is the history of man's struggle to wrest a living from its resources. It is thus relevant not only to current interests in social evolution but also to the theme of this symposium.

The first settlers in Cumbria, the Mesolithic hunter-gatherers, were probably too sparse and ill-equipped to have much impact on the mature forest but, from the beginning of the Neolithic, when the Langdale stone axe was perfected, the pollen record shows evidence of selective tree-felling for fodder and clearance for grazing and cultivation. As the mountain grazings were cleared, we see the emergence of a pattern of land ownership, similar to that of Celtic Wales and southern Scotland (Barrow 1975), with land divided between watersheds into great estates, each containing an area of mountain tops and a reserve of exploitable lowland forest. Consolidated under Anglo-Saxon and Viking occupations, these secular divisions were taken over wholesale by the Norman rulers and, as we move into the period of written records, we see the development of a struggle between rulers and ruled for the forests' resources, which continued to recent times. In Cumbria this history is unique, distinguished first by the advent of the Norse settlers and later by Scots raiders who, by habitually burning all before them, created a brisk demand for building timber. In the Furness area, the demand for fuel for ore smelting devastated the forests in Elizabethan times and led to organized management of the woods, generally on a 14-year rotation, as coppice-with-standards. Today, the economic potential of Cumbria's woodlands seems so low as to be less relevant to their management than scientific and cultural values. It is not too late in the day to recognize that Cumbria's surviving ancient woodlands are a rich repository of Cumbrian history and a legacy we ought to conserve.

2.2 The elm decline and the Neolithic clearances

The history of forest resource utilization starts in Cumbria in the Neolithic, at the beginning of the sub-Boreal period, when a decline in elm (*Ulmus* spp.) pollen appears in all pollen diagrams from NW Europe (Satchell 1983). Carbon-dating puts this decline in the Lake District between 3390 BC and 3150 BC. It is still controversial whether it resulted from selective utilization of elms by early farmers or whether a continent-wide pathogen, comparable with Dutch elm disease, was responsible, but the generally accepted explanation

(Troels-Smith 1960) is that a new technique of keeping stalled livestock was then introduced into Europe on a very wide scale and that these animals were fed mainly by gathering leafy branches of elm. Coinciding with the elm decline, ribwort plantain (*Plantago lanceolata*) appears for the first time, together with other weed species – nettle (*Urtica* spp.), bracken (*Pteridium aquilinum*), grasses (Gramineae), docks (*Rumex* spp.) and mugwort (*Artemesia* spp.). Pollarding, the frequent cutting of upper branches, is believed to have led to the selective destruction of elm trees and then to the development of what is known as the 'Landnam', a system of primitive shifting cultivation in which all trees were felled. This was followed by invasion of weeds and bracken which were succeeded by birch (*Betula* spp.) and other pioneers, and then by a return to high forest. At certain sites the elm decline is very clearly associated with evidence of Neolithic activity – at Blea Tarn clearance of upland forest, at Barfield Tarn clearance of lowland oak (*Quercus* spp.) forest for cereal growing – and at all sites it is associated with evidence for increased runoff and erosion of mineral soils (Pennington 1981).

It has been objected that the population of Neolithic man was too small to have much effect on the forest, but experiments in Denmark showed that three men with stone axes could clear about 500 m² of forest in four hours. In Cumbria, the long-bladed axe-heads from the Great Langdale, Glaramara and Scafell factories were the instruments of forest clearance, and the major exploitation of rock outcrops at these sites testifies to the scale of clearance both locally and regionally. Evidence for severe soil erosion following Neolithic tree-felling comes from several Cumbrian lakes. A sediment core from Angle Tarn, not far from the Langdale axe factory, shows successive inwash of organic soil, distinguished by its high carbon content, then weathered mineral soil, distinguished by high sodium and potassium, and lastly unweathered mineral soil, comparatively rich in calcium as well as sodium and potassium (Tutin 1969). Such stratigraphic changes are found only in lakes where the pollen evidence demonstrates partial forest clearance at the elm decline; and the distribution of these lake basins coincides with the distribution of the early Neolithic population. The evidence for soil erosion following human interference with the primary forest is therefore strong (Vincent 1985).

Once the brown earths of the forest were eroded, trees had difficulty in regenerating, and much of the forest in these cleared areas was replaced by shallow-rooting and less-demanding grassland, heath and peat communities. At Red Tarn Moss near Wrynose Pass, a layer of buried birch wood, carbon-dated at 1940 BC±90 lies beneath a layer of *Sphagnum* peat (Pennington 1965). The buried birch wood, like other buried birch forests of the Lake District mountains above 360 m, is believed to indicate accelerated peat formation caused by a combination of factors – leaching and erosion initiated by forest clearances and high grazing pressure, poor soil and high rainfall. In many areas the forest never returned, but on base-rich soils, for example around Morecambe Bay, the secondary forest which regenerated after the Landnam episodes remained relatively undisturbed until Viking times.

2.3   The Bronze Age to the Vikings

The pattern of small temporary clearances continued in Cumbria throughout the Bronze Age, which began about 1700 BC. The pollen record from Seathwaite Tarn, a valley lake with a mountain catchment, shows a fall in oak pollen from 20% to 5% during a period carbon-dated between about 2050 BC and 1080 BC. The oak was replaced by heather (*Calluna vulgaris*) and grass pollen, undoubtedly the result of occupation of these uplands by populations responsible for the numerous cairns and circles of Bronze Age type. Both at Seathwaite Tarn and at Devoke Water, these changes in the pollen record are associated with a striking fall in the carbon content of the sediments. At this date the climate appears to have been milder and drier, favouring nomadic animal husbandry, and it is believed that the severe reduction in the upland oak forest and the subsequent erosion reflect the activities of this pastoral economy. The prevailing treelessness of the uplands of the Lake District must date from these Bronze Age clearances.

At the end of the prehistoric period, there was a general expansion of agriculture. On Stainmore scrub woodland was cleared about 50 BC (T Clare pers. comm.), and on the Howgills a transition from forest to open moorland vegetation has been tentatively attributed to Iron Age/Romano-British times (Harvey, Oldfield & Baron 1981). In a number of upland tarns, retardation layers indicating a drier climate are followed by changes in the pollen record showing an expansion of forest clearance. This 'Brigantian' (Pennington 1974) clearance correlates well with a retardation layer in Helsington Moss (Smith 1959), dated about AD 400,

and evidence from Devoke Water of cereal cultivation about AD 200. This Romano-British episode in farming history had a marked impact on parts of Cumbria, as shown by a sharp decrease in tree pollen from Coniston Water and Ennerdale Water. Many upland farms of this period are now being discovered.

Early in the 7th century a movement of Anglian peoples from Northumbria through the Tyne gap and over Stainmore into the upper Eden valley began a new phase of conquest and settlement, which by the end of the century had occupied much of the better agricultural land in the lower Eden, Kent and Lune valleys, Low Furness, the north-west Cumbria coastal strip and the Solway plain (Fell 1973). Surviving artefacts from these people are few in Cumbria – fragments of stone crosses, Northumbrian coins, grave goods, museum specimens of doubtful provenance and, from a much later date in the 11th century, a document, Gospatric's writ. Their place names, however, are numerous and widely scattered, often a topographical place name ending in 'ing' and signifying a group of people dwelling in an area. Combined with the Old English suffixes 'ham' (estate or homestead) or 'tun' (farmstead, village), such names as Aldingham, Hensingham, Helsington and Killington occur in south and west Cumbria and Low Furness. Many place names emphasize that these people were dairy farmers – Sedgwick (Siega's dairy farm), Cunswick (the king's dairy farm), Keswick (cese, cheese), Butterwick (butere, butter), running mixed farms on the deeper drift and calcareous soils best suited for arable cultivation. A pair of iron shears found in an early burial at Wiseber Hall, Kirkby Stephen, suggests that these people may also have kept sheep, but in general they were lowland farmers, not extending into the upland grazings over 200 m. Rydal (ryge, dael – the valley where rye was grown) and the Old English 'mere', a lake, in Windermere, Grasmere, Buttermere show that Anglian settlements penetrated into the main Lake District valleys. Their cattle will have grazed the valley woodlands, timber will have been felled for houses, farm buildings and implements, and land will have been taken for cultivation, extending the clearings in the lowland forests of their British predecessors.

Another phase of woodland clearance, high in the pollen profile at about AD 1000 appears to be associated with Viking place names (eg Blelham Tarn) and the establishment, on rising ground along the Lakeland foothills, of new farmsteads in clearings with 'thwaite' names (eg Thackthwaite, Thornthwaite) (Winchester

1985). Pig keeping appears to have been an important part of their economy and many former shielings have swine names – there are at least seven Swinesides and Swinesetts in the area (Whyte 1985). In the Howgills, overgrazing by Viking sheep is believed to have led first to destruction of the vegetation, and then to major gullying and the formation of some of the small steep-sided valleys which are such a feature of these hills today (Harvey et al. 1981). These Viking subsistence farmers had a profound effect on the ecology of the Lake District as they cleared the forests around their settlements. From about AD 900–1000, the rate of sediment accumulation in the lowland lakes increased four-fold as the soil washed off the ploughed hillsides.

2.4   Forests and free chases

By the 13th century, land settlement in northern England had evolved into a system of land tenure in which the upland pastures were divided between groups of communities centred on lordly seats in the adjacent lowlands. The upland element, covering the estate's portion of the Lakeland massif, was retained by the lord as a private 'forest' or, strictly, free chase. In the lowland component, the lord retained direct control over relatively few settlements, mostly those near his seat, the majority of communities or vills being freehold estates bound by dues and services to the overlord at his court.

By analogy with Yorkshire and Northumbria, this pattern of land holding in Cumbria is believed to be of great antiquity, perhaps of Celtic origin but certainly pre-Norse. When the Norse invaders arrived, they appear to have taken over the seats of power of the English overlords and so gained access to the resources of the whole of their estates (Winchester 1985). When the Normans arrived, under this hypothesis, the system continued with the same structure, Norman barons replacing Norse overlords.

In Cumberland north of the Derwent, William Rufus retained the free chase as the royal forest of Inglewood; south of it he gave the forest of Copeland and a number of other free chases as private forests to the great magnates who had helped him to secure his northern borders. The royal forests were administered under a stringent system of rules known as the 'forest law', outside the ordinary law of the land. It was the subjecting of land to this law that constituted a forest, not its vegetation, for a forest might include open country. Indeed, the word originally had nothing to do with trees.

In many respects there was little difference between life in a royal forest and in a private forest. Except under licence, a man living within a private forest might not 'assart' (take new land into cultivation), or enclose land, or build, hedge, ditch, drain, cut down trees or even collect firewood, except under the eye of the forester. He could not hunt deer, wolf, boar or even the smaller animals, or cut down a bush which might give food or shelter to a deer. Restrictive though these rules were, they were tightened up even further by Henry II. It was already illegal to kill a deer, but under Henry it was forbidden to have dogs or bows and arrows in the forest without a warrant. These forest regulations acted as a damper on agricultural improvement and in the 12th and 13th centuries, a period of expanding population, were felt to be particularly repressive.

By the late 13th century, a considerable difference seems to have developed between the management of Inglewood and that of the private forests. Basically, the Crown's concern for Inglewood was to maximize profit, with a minimum of administrative cost, while preserving the game and its habitat. In practice, this meant discouraging permanent settlement in surviving unsettled areas but actively encouraging the inhabitants of neighbouring communities to 'agist' (graze) their stocks in the 'launds' (glades) and named pastures in the forest. By contrast, the private forests were seen simply as a valuable upland resource, to be exploited as profitably as possible without excessive concern for game preservation. Their owners actively encouraged peasant settlement, granted away large blocks of forest land to monastic houses, and established their own 'vaccaries' (large cattle farms) in the forests (A Winchester pers. comm.).

The hard commerce of forest resource management in Inglewood is illustrated by a series of annual statements prepared by the Keeper of the Forest (Parker 1909), which show that the greater part of the revenue was from the proceeds of grazing lands within the forest. At the beginning of the reign of Edward III, £36 was realized from this source, more than half the total income, and the proportion increased as time went on.

Timber was both sold from the forest and supplied without charge for many major disasters and for defence. In the 13th and 14th centuries, three fires occurred in Carlisle and the manors were overrun by the Scots who burnt and destroyed all they could. The King issued a series of writs to the Keeper of the Forest instructing him to supply oaks to repair the damage and the fortifications. Some of the more interesting notices include a grant to the Bishop of Carlisle of 50 oaks, to be chosen at wide intervals so as to do least harm to the forest – a proviso suggesting that Inglewood timber production was carefully managed. In 1295, the Justice of the Forest was ordered to deliver to the King's engineer as many oaks as he should choose, to make engines for Carlisle castle, and in 1323 the constable of Carlisle castle was given as many oaks as he needed to construct wooden pele towers at places where the walls needed repair, until they could be rebuilt in stone.

Many other grants of oak timber were made to repair damage by fire; for example, 20 oaks were provided in 1294 to the prior of St Mary's, Carlisle, to repair his church which had been accidentally burnt. Much greater damage was done by the great fire of Carlisle in 1391. The Patent roll for that year shows the amount of timber needed for repair: 'Grant to the citizens of Carlisle on their petition, alleging that buildings to the number of 1500 had been burnt in their principal streets, Castlegate, Ricardgate and Bochardgate, and the market place: relief for four years of the farm (ground rent) of £80, and 500 oaks in the forest of Inglewood, dry at the top, for rebuilding'.

Omitting the disturbed period of Edward II's reign, figures for the sale of timber from Inglewood begin with the second year of Edward III. They amount initially to about £6 a year, declining to about a third of this in the next 25 years. Taking into account the grants of grazing lands which were being made at this period, it appears that Inglewood was being stripped of timber from about 1340.

From time to time, substantial sums were realized by the sale of bark for tanning (Table 1).

Table 1. Revenue from the sale of bark for tanning from Inglewood

		£	s	d
1361	From William de Stapleton for the tan of the forest	10	6	8
1377	Sale of bark in Penrith ward	2	6	8
	Sale of bark in Gaitsgill ward	3	10	0
1402	Sale of bark in the ward of Carlisle John Kardoil and his friends	4	0	0

Accounts were given each year of the sums obtained from the sale of dead and fallen wood. Thus, in 1331, there is 16s 8d for the sale of wood blown down by the wind, and in 1344 9s from the same source and 14s from dead wood lying on the grass in the covert of Plumpton. Sometimes it is noted that the dead wood is for carts or for charcoal burning. The right to make charcoal from dead wood was also claimed by the priory of Carlisle from the time of Henry I (Parker 1910), but there is no evidence of any organized charcoal industry in these reigns.

Pannage, the right to turn out pigs to feed in the forest, produced substantial sums, nearly £6 annually on average, during 15 years in the reign of Henry III. Smaller sums were made from fishing rights, dog licences and fines levied on stock found straying in the forest. Sales of unclaimed strays provide a useful guide to the value of these revenues; for example, in 1344, three plough horses fetched 10s, 4s, and 2s 6d, a heifer 4s and pigs 6d a head.

Purprestures, encroachments on the forest by way of building or enclosure, raised small sums for licences and rents and unlicensed enclosures are frequently the subject of court action. In 1285, on the day following All Souls, a series of court hearings concerning Inglewood began, the cases being recorded in a great roll of Pleas of the Forest. It provides many examples of purpresture, mostly arable enclosures of less than two acres but some concerning sheepwalks. Three refer to earlier years:
'Robert, bishop of Carlisle, deceased, made a sheep walk in Mikilgil fourteen years back, without warrant.'
'John de Capella made a sheepwalk at Mikilgil in the King's forest.'
'Henry de Goldington, deceased, made two houses in the King's forest nine years back and a sheepwalk 16 years back.'

Taken in conjunction with the substantial revenues from grazing rights, these cases point to considerable pressure on the royal forest for development, and many similar examples are known from contemporary records of private forests. When it appeared that the creation of an assart or enclosure would benefit the King or the lord of the private forest, permission would be given, the hunting being reserved for the lord. Thus, in the forest of Copeland between 1200 and 1213 Richard de Lucy granted one Reginald and his heirs the 'common right of Braystones that they may freely assart and build within their right divisions, saving to me and my heirs hart and hind, wild boar and sow and hawk when any shall be there'. This is obviously a licence to break up the waste.

The rights granted to Calder Abbey in the middle of the 13th century include the right to 'cut down and prostrate the branches of trees throughout all the woods of Copeland Forest for the feeding of animals in winter to the first of May'. The deer's main winter sustenance was the trees of the forest, and the grant of rights to lop branches for feeding livestock shows how far the lords of Copeland had moved from the protection of their deer to land development. An increasing income from cultivated land overcame the joys of the chase, enclosure proceeded, the deer gradually retreated and the forest boundaries were reduced. A study by Liddell (1966) shows how the forests of Copeland and Derwent Fells decreased from the 12th century to about a third of their size in the 16th century (Figure 1).

2.5   Manorial custom and the rise of the coppice system

Greenhew and houseboot

The practice of lopping tree branches for stock to browse, begun in the Neolithic and documented as a right of the monks of Calder Abbey in the 13th century, is recorded again in the manor of Windermere in the court rolls for 1416: 'What Man or Woman dwelling within the Forest that felleth any wood for Fewell in the Wood assigned to his Neighbour for Cropping but Ellers and Birks (other than alder and birch) forfeits 3s 4d'. In Elizabeth's reign, it is explained again in a decree concerning the woodlands around Hawkshead '. . . the custumarie tenntes . . . for the relief and necessarie sustac'on of their beasts, shepp' and other catall have alwaies heretofore tyme out of the rem'embrance of man yearly used and yet yearly do use to fall and cutt the underwoods and to shred lopp' topp' cropp' and bruse (browse) all other woodes and trees . . .' (Fell 1908).

Underwoods, as distinct from woods of warrant, are defined for the Barony of Kendal by Isaac Gilpin, steward to the Levens estate, in a manuscript written about 1650–60 (Bagot 1962): 'Underwoods are commonly hazels, willows, alders, thorns and the like; woods of warrant are oak, ash, holly and crabtree'. In Sir Daniel Fleming's court at Withburne, near Rydal, these were defined slightly differently: underwoods included hazel (*Corylus avellana*), willow (*Salix* spp.), elm, rowan (*Sorbus aucuparia*), yew (*Taxus baccata*)

and alder; woods of warrant included birch and hawthorn (*Crataegus monogyna*) (Armitt 1916). In practice, the distinction appears to have been between timber and non-timber trees, as implied by an exchange (Jones 1971) in the parish of Witherslack. 'Suppose', asked a tenant, 'I plant an acre of ground with acorns, have I not the right to cut it down when it is fit for coaling?' 'Surely', replied the lord's agent, 'if you cut it in underwood or shrubbery'.

Expanding his definition, Gilpin continues. 'Tenants commonly claim houseboot, hedgeboot, plowboot and cartboot out of woods growing on their tenements but they cannot cut down any woods of warrant without the warrant or licence of the lord or his bailiff. Nor have they any property rights in the woods but only a duty or acknowledgement which is called greenhew which is one penny or in some places twopence to be paid yearly at the Lord's court. By virtue of this duty of greenhew tenants claim liberty of cutting underwoods in their own tenements and also lopping or cropping woods of warrant growing in their own tenements. Unreasonable lopping, cutting off branches too near the bole or cutting off the top of the tree constitutes trespass as does unseasonable lopping, cutting the trees when the sap is up or rising or descending.'

This last regulation was apparently not applied in all the manors of the Barony of Kendal. A record from Windermere dated 1630 states that 'Whosoever doth cutt down or breake any other Men's Ash leaves forfeits 1s' (Armitt 1906), a clear indication that ash branches were sometimes cropped in leaf. Father West describes the custom in High Furness in 1774: 'The woodlanders . . . were charged with the care of the flocks and herds which pastured . . . the fells . . . and in winter to browse them with the tender sprigs of the hollies and ash'. The practice is recorded by several later authors (Forrester 1985).

Houseboot, the custom of the lord providing timber for repairing the tenants' houses, was not, in Gilpin's view, a right. Rights to timber are the lord's, he writes, and tenants cannot have timber out of the woods on their own tenements, still less from the lord's woods, unless the lord 'of his own Good will do Give it'. Grudgingly acknowledging the custom, he continues 'If the tenant would build larger houses for his own pleasure or convenience rather than for necessity it were an unreasonable custome that the landlord should find his Tennant tymber for building for his pleasure or vain Glory'.

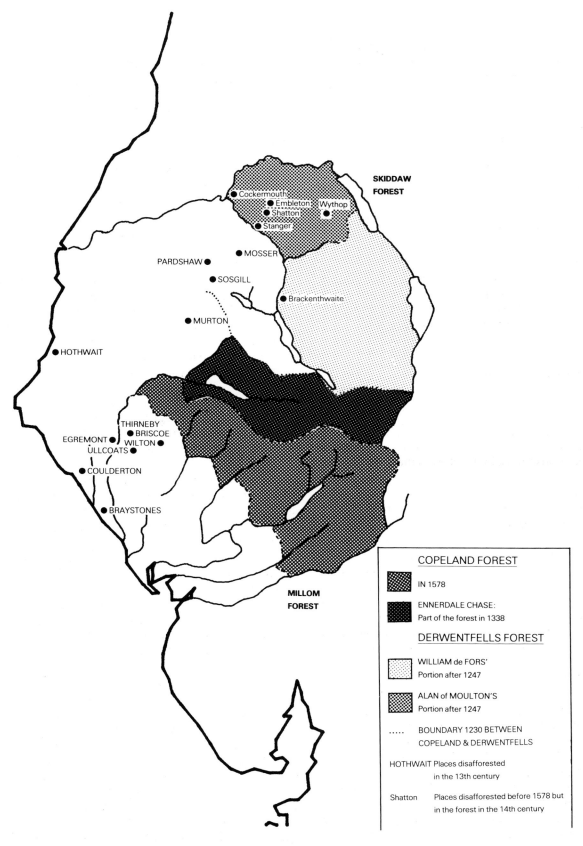

SKIDDAW
FOREST

● Cockermouth
● Embleton   Wythop
● Shatton
● Stanger

PARDSHAW ●   ● MOSSER

● SOSGILL

● Brackenthwaite

● MURTON

● HOTHWAIT

THIRNEBY
● BRISCOE
EGREMONT ● ● WILTON
ULLCOATS ●

● COULDERTON

● BRAYSTONES

MILLOM
FOREST

COPELAND FOREST

IN 1578

ENNERDALE CHASE:
Part of the forest in 1338

DERWENTFELLS FOREST

WILLIAM de FORS'
Portion after 1247

ALAN of MOULTON'S
Portion after 1247

..... BOUNDARY 1230 BETWEEN
COPELAND & DERWENTFELLS

HOTHWAIT Places disafforested
in the 13th century

Shatton Places disafforested before 1578 but
in the forest in the 14th century

Figure 1. The private forests of west Cumbria (Liddell 1966)

Every year at the lord's court, the tenants were required to swear on oath that they had not 'Grubbt Sold or cutt down any wood of warrant as ash oake holly or crabb within this lordship since last Court here holden without licence at the hand or delivery of his officer so help you God'. This done, they paid their greenhewes.

The charcoal industry

Charcoal burning is referred to as early as the 12th century in several documents in the register of the great Cistercian abbey of Holm Cultrum. William, Earl of Albermarle, who died in 1179, granted a forge in Whinfell, probably Smithy Fell, west of Lorton, 'with green and dry wood to make charcoal'. The operation of the smithy and smelting iron ore are specifically distinguished in the early 14th century by Antony de Lucy, Sheriff of Cumberland; 'Antony de Lucy grants to Holm abbey leave to take, carry and burn dead wood in his wood of Allerdale, to make charcoal for their forges in the Holm; but not to take charcoal in the said wood to smelt their iron-ore without special leave' (Grainger & Collingwood 1929).

Again, an undated inquisition towards the close of Henry III's reign records that Thomas de Multon, a hereditary warden of the forest of Carlisle, and his predecessors 'had been accustomed to make charcoal of the loppings and other timber which belongs to him by reason of this bailiwick, wherever they wished in the said Forest' (Parker 1905).

Furness Abbey also had the right to take wood for smelting. In 1537, at the time of the dissolution of the monastery, two lists of property and rents were compiled, one by the Abbot, the other by the King's Commissioners. Both record that the right to maintain three smithies, formerly enjoyed by the monks, had been let to John Sawrey and William Sands. The Commissioners' list (Beck 1844) includes the following. 'Also there is moche wood growing in Furness fells in the mounteynes there, as Byrk, Holey, Asshe, Ellers, Lyng, lytell short Okes and other Undrewood, but no timber of any valewe, wherein the Abbots of the same late Monastery have been accustomed to have a Smythey, and sometime two or three, kepte for making of Yron to th' use of their Monastery. And so nowe the Commyssyoners have letten unto William Sandes and John Sawrey as moche of the said woodes, that is to say of Byrkes, Ellers, Hasells, old rotten trees and other underwoods, as might maynteyne iij (three) Smythys, for the which they ar content and agreed to paye yere-

ly to the Kings Highness, as long as it shall please his grace they shall occupy the same, XX li (£20).'

In the following year, 1538, a court was held at Hawkshead at which fines were levied on William Blumer of Satterthwaite and others who had made charcoal and 'used the smiths' art'. The demands of the bloomsmithies so destroyed the woods that in 1564 Queen Elizabeth closed three recently erected smithies in Hawkshead, by decree, to protect the livelihoods of her other tenants: 'The woodes be sore decayed and dailie more and more are like to fall into great decaye not only by reason of certain iron smithies . . . but also for that the customarie tenntes . . . as well as for their proper fuell and for mainten'nce of their hedges and other necessaries . . . yearly doe . . . cutt the under-woodes and lopp . . . all other woodes and trees . . .'. The tenants were allowed to make charcoal and smelt iron for their own use, using, as in earlier reigns, 'the shreadings, tops, lops, crops and underwood, but not timber'.

In 1557, in the whole of High Furness, there were only 1280 timber trees and 8260 saplings. Only three years later, however, a decree mentions that woodland was sufficiently plentiful in the Furness Fells to enable 160 acres to be set aside at Garthwaite, 60 acres at Elterwater Park and 60 acres in Satterthwaite and Grizedale to supply charcoal to the company of the Mines Royal (Fell 1908).

The two surveys, of 1538 and 1567, appear to span the beginnings of managed coppice, in agreement with the mid-16th century date given by Oldfield (1965) from pollen analysis of lowland deposits in south Cumbria. Coppicing provided small-diameter poles by frequent cutting of trees, close to ground level. Coppice was often interspersed with trees which were allowed to grow to maturity, known as standards. The data are also consistent with Isaac Gilpin's description in the 1650s of coppicing as 'ancient': 'the Landlord may and doth usually make Springs or Coppises of the Tennants Grounds as have been aunciently copised or sprung after the Lord have sufficiently fenced them'. A Parliamentary survey carried out in 1649 shows the woods much recovered from the devastation of the previous century but with the standards still immature. Growing on the lands of the customary tenants of High Furness were 'between three and four thousand Timber Trees (most of them but of small growth)'.

The rights allowed to tenants, in return for their greenhew penny, changed with changing woodland economics. Two leases from the Low Meathop estate, near Grange-over-Sands, provide examples. The first, recorded for 1684 in an estate minute book, reads 'Andrew Muggell had his lease for seven years, one year at £20, and the other six years at £25 per annum, with liberty to cut coal wood and others, crab and holly excepted, and to sell a hundred of their cartloads of turf, oaks and ashes reserved to the feoffees, with liberty to cut and carry them away. House boote and plough boote to be allowed him at the discretion of the feoffees . . .'. He was thus allowed to sell peat, to make charcoal and to have such timber for buildings and implements as the landlord thought fit.

The next lease to Thomas Barrow in 1690 makes an important change in reserving to the landlord the right to make charcoal. 'In consideration of a yearly rack rent of £26 10s 0d he is to enjoy the messuage of Nether Mythop with its appurtenances . . . not delving any in the enclosed ground, excepting all trees and shrubs or underwood to his landlord's (use) with free liberty . . . . To sell and cut down, cord, coal and cut or dig up earth to cover the same, burn into charcoal . . . and dispose of it at their pleasure, . . . he shall not sell or give away any oakes, ashes or underwood . . . he shall preserve the wood from spoil or hurt of cattle, and all other harm or destruction.' On the neighbouring estate of the Manor of Witherslack, the tenants leased the timber until 1734 (Jones 1971), but in all subsequent leases the right to the timber and to make charcoal was retained by the landlord, the value of the coppice crop increasing at every rotation. By the mid-18th century, charcoal was in such demand that, according to a handbill of 1748, the iron masters had formed a combination to standardize the measure of charcoal and fix its price, and four of the biggest woodland property owners, 'Gentlemen of the County', had planned a secret agreement to form a charcoal monopoly. New land was planted for coppice, Stockdale Coppice on Meathop Fell providing an example. In an estate inventory of 1760, it is classed as woodland but called simply 'part of Meathop Fell'. In 1773 it was called 'the new coppy part of the Fell', the managers' report stating that 'the young spontaneous birch plants on the new woodlands on the fell are flourishing'. In 1781 they found that the woods in 'the new enclosure on the Fell' were 'short and shroggy' . . . and ordered 10 000 young birch plants to 'fill up the vacancies'. From the conservation viewpoint, it is interesting that this wood today shows the same botanical composition as adjacent ancient woodland (A H F Brown pers. comm.).

Other woodland products

A number of by-products of coppice woodland contributed to its value (Table 2). Hoop sticks, thin poles of

Table 2. The value of produce from Lime Garth Wood, adjacent to Stockdale Coppice on Meathop Fell, 1789 (Satchell 1984)

	£	s	d
110 dozen sacks of charcoal at 23s per dozen	126	10	0
130 quarters long cut bark	65	0	0
Hoop sticks	7	10	0
Coard wood	6	10	0

oak and hazel, were not only used for local cooperage but were exported from south Cumbria in large quantities to Manchester and Liverpool, particularly for making the large casks used in the coal export and sugar import trade with the West Indies. Coard wood in 1789 was probably firewood but at later dates the term is also applied to wood for bobbin making.

In the early 19th century, oak bark for tanning increased greatly in value (Table 3). The list gives an in-

Table 3. The value of produce obtained from Town Head Wood near Finsthwaite by a wood merchant, 1815–17 (Satchell 1984)

	£	s	d
Bark	656	5	2½
Charcoal	212	8	11
Hoops	128	17	4
Oak hoops	26	12	0
Mast hoops	13	0	6
Sap spokes	57	7	0
Birch and alder poles	48	18	5
Ash poles	12	18	10
Smarts	24	12	0
Rods	20	9	6
Prop wood	15	18	1
Swill wood	11	3	5
Powder wood	5	2	0
Seal wood	2	5	4
Scythe poles and brush sticks	1	17	11

dication of the diverse industries — tanning, smelting, mining, quarrying, agriculture, shipping and woodland crafts — which used woodland produce. Some of the woodland crafts were partly carried out in the woods themselves. For example, making swills, which were baskets for general farm use or, in specialized forms, for coal and iron ore, utilized 'rods' and 'seal wood'. Sallow or willow formed the frames on to which the 'smarts' (split oak laths) were woven. The importance of these woodland trades in the life of Cumbria's rural communities is illustrated by a list of 52 craftsmen in the parish of Witherslack compiled by Jones (1971) for the year 1786. Approximately half of them are concerned with wood, viz six charcoal burners, five chair makers, four carpenters, four hoop makers/coopers, two candle box makers, one wood cutter, one turner, one clogger, one basket maker.

Decline

In the 19th and early 20th centuries, the modernization of the tanning industry and the replacement of charcoal by coke resulted in a decline in the value of coppice woodland, which is well illustrated in Table 4.

Table 4.   Prices obtained at auction for woodlands in Low Meathop and Witherslack (Satchell 1984)

		£
1808	Limegarth	443
1834	Limegarth	400
1849	Limegarth + Stock Dub	260
1863	Limegarth + Stock Dub	216
1876	Limegarth + Stock Dub + Halecat	214
1892	Limegarth + Stock Dub + Halecat + Boundary Woods + Adams Wood + Long Dale	123

The management of Cumbria's woodlands in the 19th and 20th centuries, discussed by Voysey (see page 18), covers a period of new markets, new management techniques and new perceptions of values and priorities. These are the foundations of modern concepts of the countryside as a national resource.

2.6   Conclusion

Although this survey of the history of man's use of Cumbrian woodlands is necessarily brief and superficial, it reveals the possibility of analysis at several levels. First, we see how such great national and international events as the Norman conquest or the dissolution of the monasteries not only changed the ownership of the forests but penetrated the social fabric to the level of the cottager picking up sticks for his fire. Then, at the regional level, we see how such local circumstances as the proximity of the Scottish border created particular local demands on the forests, or how the existence of the nearby seaport of Liverpool created a niche for the woodlanders to obtain a living

from weaving coal baskets. Third, at the parish level, as at Witherslack, we see how approximately half the craftsmen were occupied in working woodland products. A vast reserve of unread manuscripts in the county's archives offices and muniment rooms awaits research to fill out this sketch.

Planning legislation acknowledges both the landscape value of woodlands and their value for wildlife conservation. Well-established procedures ensure that both are considered in the preparation of National Park plans. The historical interest of woodlands has never been recognized as a separate and specific cultural value, although in various acts of Parliament, concerned with historic buildings and archaeological sites, it is long established as a criterion of merit, allied to but distinct from architectural interest. Many of our woodlands are much older even than our parish churches and are no less part of our heritage. In presenting this first paper on Cumbria's woodland resources, I would like to suggest that any fragments of old woodland that still survive in the county should be viewed not only as landscape and conservation resources, but as part of the cultural inheritance we hold in trust for our successors.

References

Armitt, M.L. 1906. Ambleside town and chapel: some contributions towards their history. *Transactions of the Cumberland and Westmorland Antiquarian and Archaeological Society*, **6**, 1–96.

Armitt, M.L. 1916. *Rydal*, edited by W.F. Rawnsley. Kendal: Titus Wilson.

Bagot, A. 1962. Mr. Gilpin and manorial customs. *Transactions of the Cumberland and Westmorland Antiquarian and Archaeological Society*, **62**, 224–245.

Barrow, G.W.S. 1975. The pattern of lordship and feudal settlement in Cumbria. *Journal of Medieval History*, **1**, 117–138.

Beck, T.A. 1844. *Annales Furnesienses. History and antiquities of the Abbey of Furness*. London.

Fell, A. 1908. *The early iron industry of Furness and district*. London: Cass (reprinted 1968).

Fell, C.I. 1973. Dark Age to Viking times. In: *The Lake District*, edited by W.H. Pearsall & W. Pennington, 236–249. London: Collins.

Forrester, R.M. 1985. Ash as fodder. *Lakeland Gardener*, **VII** (2), 18–24, and **VII** (3), 37–38.

Grainger, F. & Collingwood, W.G. 1929. *The register and records of Holm Cultram*. (Cumberland and Westmorland Antiquarian and Archaeological Society record series vol. 7.) Kendal: Titus Wilson.

Harvey, A.M., Oldfield, F. & Baron, A.F. 1981. Dating of postglacial landforms in the central Howgills. *Earth Surface Processes and Landforms*, **6**, 401–412.

Jones, G.P. 1971. *A short history of the manor and parish of Witherslack to 1850.* (Cumberland and Westmorland Antiquarian and Archaeological Society tract series 18.) Kendal: Titus Wilson.

Liddell, W.H. 1966. The private forests of S.W. Cumberland. *Transactions of the Cumberland and Westmorland Antiquarian and Archaeological Society,* **66**, 106–130.

Oldfield, F. 1965. Problems of mid-postglacial pollen zonation in part of north west England. *Journal of Ecology,* **53**, 247–260.

Parker, F.H.M. 1905. Inglewood Forest. *Transactions of the Cumberland and Westmorland Antiquarian and Archaeological Society,* **5**, 35–61.

Parker, F.H.M. 1909. Inglewood Forest. Part IV. The revenues of the Forest. *Transactions of the Cumberland and Westmorland Antiquarian and Archaeological Society,* **9**, 24–37.

Parker, F.H.M. 1910. Inglewood Forest. Part V and VI. *Transactions of the Cumberland and Westmorland Antiquarian and Archaeological Society,* **10**, 1–28.

Pennington, W. 1965. The interpretation of some post-glacial vegetation diversities at different Lake District sites. *Proceedings of the Royal Society B,* **161**, 310–23.

Pennington, W. 1974. *The history of British vegetation.* 2nd ed. London: English Universities Press.

Pennington, W. 1981. Records of a lake's life in time: the sediment. *Hydrobiologia,* **79**, 197–219.

Satchell, J.E. 1983. A history of Meathop Woods. Part 1 – Prehistory. *Transactions of the Cumberland and Westmorland Antiquarian and Archaeological Society,* **83**, 25–32.

Satchell, J.E. 1984. A history of Meathop Woods. Part 2 – The Middle Ages to the present. *Transactions of the Cumberland and Westmorland Antiquarian and Archaeological Society,* **84**, 85–98.

Smith, A.G. 1959. The mires of south-western Westmorland: stratigraphy and pollen analysis. *New Phytologist,* **58**, 105–127.

Troels-Smith, J. 1960. Ivy, mistletoe and elm: climatic indicators – fodder plants. *Danmarks geologiske undersogelse,* **4**, 4, 1–32.

Tutin, W. 1969. The usefulness of pollen analysis in interpretation of stratigraphic horizons, both Late-glacial and Post-glacial. *Mitteilungen der Internationalen Vereinigung fur theoretische und argewarate Limnologie,* **17**, 154–64.

Vincent, P. 1985. Pre-Viking changes in the Cumbrian landscape. In: *The Scandinavians in Cumbria,* edited by J.R. Baldwin & I.D. Whyte, 7–16. Edinburgh: Scottish Society for Northern Studies.

West, T. 1774. *Antiquities of Furness.* London.

Whyte. I. 1985. Shielings and the upland economy of the Lake District in medieval and early modern times. In: *The Scandinavians in Cumbria,* edited by J.R. Baldwin & I.D. Whyte, 103–117. Edinburgh: Scottish Society for Northern Studies.

Winchester, A.J.L. 1985. The multiple estate: a framework for the evolution of settlement in Anglo-Saxon and Scandinavian Cumbria. In: *The Scandinavians in Cumbria,* edited by J.R. Baldwin & I.D. Whyte, 89–101. Edinburgh: Scottish Society for Northern Studies.

The composition of woodlands in Cumbria

R G H Bunce

Institute of Terrestrial Ecology, Grange-over-Sands

3.1 Introduction

In pre-historic times, Cumbria was almost completely covered with woodland which, as described by Satchell (see page 2), was progressively and selectively cleared by man for agricultural purposes. These factors have led to patches of woodland being left on land with the lowest value for agriculture. The remaining, or ancient, woodland is often on steep slopes and shallow soils, so it is difficult to appreciate the original woodland composition across Cumbria as a whole. The area of woodland probably reached its lowest point towards the end of the 17th century, but since then it has been increasing; much of this increase has been relatively recent and is coniferous in character. The expansion of woodland cover has provided shelter for deer which have increased in the last 50 years. Whilst there are no comparable records for other woodland animals (eg birds), their numbers also have probably increased, at the expense of moorland species, as has occurred elsewhere in Britain.

The primary environmental influence in Cumbria is the climate, which is relatively mild even in winter, with high rainfall throughout the year, reaching a maximum in winter. This oceanic climate determines the overall composition of the flora and fauna in the area. Another general influence is the island status of Britain, which has restricted immigration of species, in comparison with other mountainous areas in Europe. At a local level within the county, it is usually the soil nutrient and moisture status which is critical in determining the vegetation composition, but this is overridden in upland areas by the influence of altitude, which restricts the distribution of some species.

3.2 The role of woodland in Cumbria

Cumbrian woodlands are important for many reasons, including the following.

i. **Amenity.** The woodlands of the Lake District are a vital part of the amenity of the area and are particularly important in visual terms. The old plantations around Windermere add very considerably to the scenery: it is only the large new coniferous plantations which are often considered deleterious to the landscape.

ii. **Wildlife.** Woodlands are very important for their wildlife, both animals and plants. Some important woodlands are protected as Sites of Special Scientific Interest and National Nature Reserves.

iii. **Timber resource.** A separate aspect of conservation is the use of Cumbrian woodlands as a source of timber. In the 17th and 18th centuries, the supply of timber was vital to the maintenance of an industrial base in Britain, although this is now of less importance.

3.3 Composition by area

According to the most recent Forestry Commission (FC) census (Table 1), the county has 54 450 ha of

Table 1. Area of woodland in Cumbria, by forest type, 1980 (Forestry Commission 1984)

Forest type	Area (ha)	% of total
Mainly coniferous high forest	35 340	65
Mainly broadleaved high forest	13 806	25
Coppice-with-standards	89	<1
Coppice	134	<1
Scrub	4 196	8
Cleared	885	2
Total	54 450	100

woodland, which amounts to about 8% of the county, whereas Great Britain overall has approximately 9% woodland. Cumbria is, therefore, comparable to the rest of Britain, rather than to Europe, where only Holland and Eire have less forest. The ownership and pattern of management in Cumbria are similar to the rest of Britain, although there is probably a higher proportion of coppice, as opposed to high forest, than in many counties. In the past, the practice of coppicing was very much more common in Cumbria than it is now.

Table 2 gives the proportion of forest occupied by the principal species. Sitka spruce (*Picea sitchensis*) is now the most important component of the conifer high forest, largely due to recent planting, whilst the traditional coniferous plantation tree was Scots pine (*Pinus sylvestris*). Oak (*Quercus* spp.) predominates in the broadleaved high forest, but it may be surprising that ash (*Fraxinus excelsior*) forms such a large proportion; this reflects the importance of the limestone areas in the south of the county. Sycamore (*Acer pseudoplatanus*) and beech (*Fagus sylvatica*), both of which are introduced, also cover large areas and play a significant part in the ecology of the county. The scrub category is dominated by birch (*Betula* spp.), whereas in the 1947 census it was mainly oak, reflecting a change in definition made by the FC between the two dates. In 1947, conifers occupied only 56% of the overall woodland area of Cumbria, whereas by 1980

Table 2. Area of high forest in Cumbria, by principal species, 1980 (Forestry Commission 1984)

Species		Area (ha)	% of total
Scots pine	*Pinus sylvestris*	4 452	9.1
Corsican pine	*Pinus nigra*	315	0.6
Lodgepole pine	*Pinus contorta*	1 837	3.7
Sitka spruce	*Picea sitchensis*	16 856	34.3
Norway spruce	*Picea abies*	4 020	8.2
European larch	*Larix decidua*	1 621	3.3
Japanese and hybrid larch	*Larix kaempferi* and *Larix × eurolepis*	3 148	6.4
Douglas fir	*Pseudotsuga menziesii*	654	1.3
Other conifers		885	1.8
Mixed conifers		1 561	3.2
Total conifers		35 349	71.9
Oak	*Quercus* spp.	3 892	7.9
Beech	*Fagus sylvatica*	1 221	2.5
Sycamore	*Acer pseudoplatanus*	1 235	2.5
Ash	*Fraxinus excelsior*	1 115	2.3
Birch	*Betula* spp.	2 508	5.1
Poplar	*Populus* spp.	79	0.2
Sweet chestnut	*Castanea sativa*	5	<0.1
Elm	*Ulmus* spp.	329	0.7
Other broadleaves		906	1.8
Mixed broadleaves		2 507	5.1
Total broadleaves		13 797	28.1
Total		49 146	100.0

this percentage had increased to 72. This increase is due in part to the large new plantings of conifers, particularly in the north of the county, but also to the conversion of broadleaved woodlands to conifers.

Table 3 indicates the change from the early 19th century, when broadleaved planting predominated, through to the present day when most planting is

Table 3. Area (ha) of high forest in Cumbria in 1980 by planting year class (Forestry Commission 1984)

Planting year class	Mainly coniferous	Mainly broadleaved	Total
Pre–1861	30	808	838
1861–1900	485	3 014	3 499
1901–1910	358	1 346	1 704
1911–1920	401	546	947
1921–1930	1 737	661	2 398
1931–1940	3 550	1 485	5 035
1941–1950	4 350	1 469	5 819
1951–1960	7 654	3 156	10 810
1961–1970	8 731	1 096	9 827
1971–1980	8 044	225	8 269
Total	35 340	13 806	49 146

of conifers, a trend which seems likely to continue. When interpreting Table 3, it is important to bear in mind that conifer rotations are shorter than broadleaved rotations. The economic pattern of recent years has led to a predominance of coniferous plantings, because of their higher financial return. On the other hand, in recent years there has been a decline in the conversion of broadleaved woodland to coniferous forest because of the high cost of management.

The Lake District Special Planning Board (1978) undertook a survey to determine changes in the area of broadleaved woodland in the Lake District National Park. A total area of 11 600 ha comprised broadleaved woodlands greater than 0.5 ha, and over 1100 ha of woodland which was broadleaved about 30 years ago is now predominantly coniferous. A further 390 ha of broadleaved woodland (over 2 ha in area) no longer exists at all. This loss is partly compensated for by 300 ha of new broadleaved woodland, although it is important to point out that this is an underestimate, because of increases in area around the margins of old woodlands. Regeneration was recorded in only 9% of the sample woodlands, and generally consisted of different species from those of the main crop. The replacement of oak by other species is likely to be part of a natural cycle, particularly as oak will not regenerate in its own shade. In addition, the charcoal industry probably favoured oak, as explained by Satchell (see page 8).

In the 1960s, much concern was expressed about the lack of tree regeneration in the Lake District. Since then, however, areas have been found where vigorous natural regeneration is occurring. There is little doubt that heavy grazing pressure is the primary limiting factor for regeneration; ungrazed sites with a wide range of soil, altitude and ground conditions have been found with regeneration. Locally other factors (eg exposure or waterlogging) may be important. The species composition of surrounding trees is an important factor in determining the type of regeneration.

The Nature Conservancy Council (Whitbread 1985) surveyed ancient woodland in Cumbria and, although there were some problems with identifying the boundaries of these woodlands, the figures confirm the decline of broadleaved woodlands. In 1920 there were 16 545 ha of ancient woodlands, and by 1985 this figure had declined by 4%, to 15 880 ha. However, only 10 518 ha of this area are now semi-natural, about 32% of the original woodlands having been planted with conifers. The loss of broadleaved woodlands is, therefore, not necessarily due to any inherent lack of regenerative capacity, but is a result of their replacement with conifers or other exotics by man. Only 2.3% of the county consists of ancient woodlands, emphasizing the scarcity of this resource.

3.4 Vegetation composition of coniferous woodland

The vegetation of coniferous plantations is often ignored by botanists, because it does not have the intense interest of traditional woodlands. Whilst these coniferous forests are, by any method of comparison, poorer than traditional oak forests, there is still much of interest in their wildlife and vegetation. A common view of coniferous woodland is of young Sitka spruce with a scattering of mosses on the ground and no other species beneath the canopy. Whilst this is true in many new forests (eg Kershope and Spadeadam), the plantations of the central Lake District are more variable, because the terrain is rocky and areas of unplantable land are scattered through the forest. However, even in the extensive forests, there are still areas where the crop has failed or where rides or tracks are present; here, relict moorland vegetation remains. On some sites in the central Lake District, the better soils allow longer rotations, and thus a mature woodland flora develops. Many of the complex woodland vegetation types seen in conifer forests in the west of North America and in Scandinavia do not occur in Britain, either because the local vegetation has not had time to adapt to the new conditions of dense shade or because suitable species have not yet evolved or been introduced.

Brown, Pearce and Robertson (1979) have shown the effects of converting traditional broadleaved woodlands to conifers, and Hill (1979) has described the changes that take place as coniferous forest develops following afforestation. The early stages of the conifer rotation are important because there is extensive growth of the existing ground vegetation, following its release from grazing pressure, providing habitats for birds such as the pheasant (*Phasianus colchicus*) and the long-eared owl (*Asio otus*). Mature conifers are also important landscape features and provide cover for wildlife. Although the blanket coverage of dense forests may generally be considered deleterious to flora and fauna, their opening up by felling increases the variability present, both in forest structure and vegetation composition.

3.5 Vegetation composition of broadleaved woodland

Maps of the distribution of individual woodland species are shortly to be produced in the *Flora of Cumbria*, but Halliday (1978) already provides a great deal of information. A wide range of purely descriptive texts is also available on the vegetation of Lake District woodlands (eg Tansley 1949; Yapp 1953). However, for the present purposes, the quantitative framework used is that of Bunce (1982), where random sample plots in woodlands throughout Britain were classified using multivariate statistical analysis. Thirty-two woodland vegetation types were defined within this classification, 22 of which are found in Cumbria.

Vegetation types 1 to 8 are characterized by lowland calcareous species, such as dog's mercury (*Mercurialis perennis*), hart's-tongue fern (*Phyllitis scolopendrium*) and spindle (*Euonymus europaeus*). Such species are at their northern British limit in south Cumbria, whereas other species of this group, eg oxlip (*Primula elatior*) and stinking iris (*Iris foetidissima*), are not found as far north as Cumbria. Thus, only three of these eight vegetation types are present in Cumbria, and these are mainly in woodlands in the south of the county. The intermediate lowland vegetation types, 9 to 16, are often associated with stream banks or enriched areas, typical species being primrose (*Primula vulgaris*), dog's mercury and bramble (*Rubus fruticosus*). These vegetation types are much more variable than types 1–8 because they also have elements from more acidic soils. Six out of eight of these types are represented, being present in the upland/lowland margins of the county. They may also be found by streamsides on the low fells (eg Rusland) or even in gorges in the uplands (eg Borrowdale). In the latter case, they are the result of local enrichment. The lowland acidic vegetation types, 17 to 24, are the most abundant in the county and are widespread throughout the majority of woodlands. They may even be present in limestone woodlands where acid drift overlays the limestone. These types include a range of cover species from bracken (*Pteridium aquilinum*) and bramble, through to wavy hair-grass (*Deschampsia flexuosa*), and even bilberry (*Vaccinium myrtillus*) and heather (*Calluna vulgaris*). Seven of these eight vegetation types are represented in Cumbria, the absent one being typical of clayey or very acidic soils in south-east England. The extreme upland types, 25 to 32, are often very variable and have a wide range of species present. Again, seven of the eight vegetation types are present in Cumbria, mainly in upland woods at the head of dales such as Borrowdale. The absent vegetation type is an extreme type, typical of exposed areas of north-west Scotland.

Cumbria, therefore, has a wide variety of woodland vegetation types, covering virtually the whole range found in Britain, from the calcareous lowlands to the very acid, exposed, upland conditions. This reflects the inherent variability of the Cumbrian environment.

Woodlands are combinations of different vegetation types and may be compared by analysing the range of vegetation types present in each (Bunce 1989). Using this approach to compare Cumbria with the rest of Britain demonstrates that Cumbria is in an intermediate position, containing elements from both north and south. However, the affinity is rather more towards Scotland than lowland England.

Two important vegetation components deserve attention.
i. **Bryophytes:** the mild moist climatic conditions have led to Cumbrian woodlands having an exceptional flora of mosses and liverworts. The Borrowdale woods, for example, contain some of the finest assemblages in western Europe. Many of these species are difficult to identify, but nevertheless are of great botanical importance.
ii. **Lichens:** some of the more sensitive species are present in Cumbria woodlands because of the relatively pollution-free atmosphere.

3.6 Cumbrian woodlands in a European context

The distribution of individual species can be used to explain the wider affinities of Cumbrian woodlands.

The bluebell (*Hyacinthoides non-scripta*) is present throughout Cumbria. However, this species is particularly important in a European context because it is found on the mild oceanic fringe of western Belgium, western France and Spain. It therefore indicates the oceanic nature of woodlands in Cumbria. Although oceanicity is the dominant climatic influence on Cumbrian woodlands, other species indicate some similarity with continental climates. Scots pine colonizes peat mosses in Cumbria as it does in Scandinavia. On the Cumbrian mosses, the ground vegetation is dominated by cross-leaved heath (*Erica tetralix*), cranberry (*Vaccinium oxycoccus*) and bilberry, whereas in Scandinavia other dominant species are found, although the vegetation has a similar form. Another example of

similarity is with woodlands of western Belgium and central Germany, but this similarity decreases as one moves progressively away from the oceanic margins. However, in France it is not until one approaches the Mediterranean, about 550 miles from the Atlantic coast, that the woodlands appear markedly different. The species composition of the canopy, which includes sessile oak (*Quercus petraea*), holly (*Ilex aquifolium*) and hazel (*Corylus avellana*), remains the same, but the ground flora species are replaced by more continental species. Further south, the Pyrenees form a major barrier to species migration, isolating the Spanish woodlands.

3.7 Sites of particular ecological significance

The most important Cumbrian woodlands in the national context are the high-altitude woods of the Lake District, the highest woodlands in England and Wales. Keskadale and Birkrigg woods lie between 400 m and 500 m above the Newlands valley, and are described by Yapp (1953). They are predominantly oak woods, low grown and windswept. Their ground vegetation is, however, more typical of upland grassland than woodland, so it is their altitudinal position which is important. Another site, which is not so well known, is at Mungrizedale and lies between 490 m and 520 m. Although similar to Birkrigg in tree composition, there is much heather and bilberry in the ground vegetation, in places only several centimetres high, its height no doubt being controlled by grazing.

Another exceptional series of sites are the coastal mosses in the south and north of the county. Here the peat has built up out of reach of the calcareous groundwater, to a situation where the soil is pure peat and receives nutrients only from rainfall. The ground vegetation, therefore, consists of species which favour very acid soils. Around the margins of these mosses, Scots pine was planted in Victorian times and is now colonizing the bog surfaces. As the tree roots dry out the bog, pine and also birch spread aggressively into the bog centre, modifying the heath vegetation until it becomes woodland.

Throughout the central Lake District, there are fragments of lake margin woodland, which are best preserved where there has been little recreational use. Elterwater has one of the best examples of marginal woodland, with birch, alder (*Alnus glutinosa*) and willow (*Salix* spp.), leading to yellow iris (*Iris pseudacorus*) and canary grass (*Phalaris canariensis*) nearer the water. Another good example is on the north shore of Bassenthwaite Lake. These remnants represent a woodland type which was formerly extensive, before man cleared and drained such areas for agriculture.

Other important woodlands virtually unaffected by man's activity are the gill woodlands, which may reach 600 m above sea level. At the valley floor, the gills have typical oak woodland species, such as wood sage (*Teucrium scorodonia*), buckler-fern (*Dryopteris dilatata*) and male-fern (*Dryopteris filix-mas*). With increasing altitude, oak declines gradually and is replaced by birch, rowan (*Sorbus aucuparia*) and willow. The ground flora also changes, to species more characteristic of open birch woodland of the north-west of Scotland, such as bog asphodel (*Narthecium ossifragum*) and yellow saxifrage (*Saxifraga aizoides*). A study recently carried out in Langdale (R G H Bunce unpublished) shows that the boundaries of these two types of woodland probably follow the original natural boundary between oak and birch. The gills thus represent a relict flora which can be used to indicate the old boundaries in the original woodland cover. Comparable vegetation is seen on the high cliffs.

Finally, there is the dwarf shrub heath of the higher mountains, composed of shrubs such as juniper (*Juniperus communis*) and willow, which represents the altitudinal limit of woodland vegetation in the Lake District.

3.8 The future

The future development in woodlands is not easy to predict. The control of planting in many areas in the Lake District, through the influence of the National Park Authority, prevents the establishment of any large new areas, at least in the central Lake District. In this area, small amenity plantings, as are currently being made by the National Trust in Langdale, are likely to be found.

In north Cumbria, large new plantations are likely to continue the trend of the last 20 years. With increasing pressure on the world's forests, several other areas around the National Park are likely candidates for these new plantations, such as the Shap Fells and the foothills of the Pennines. Elsewhere, the current concern about the excess of agricultural land may well lead to small local plantings, but this is unlikely to be a significant factor in the county as a whole, in terms of area, although it may well affect the scenery.

Conservation will continue to be an important issue. Upland sites where the lack of regeneration through grazing is a problem contrast with the lowland peat mosses where the colonization by trees is removing heather from the ground vegetation. In these cases, management objectives need to be clearly defined before an appropriate course of action is adopted.

The maintenance of cultural artefacts could also be an important local influence on woodland management. Considerable interest has recently been expressed about the way in which woodlands were used in the past and in the re-introduction of ancient industries, such as charcoal burning and the making of swills.

3.9 Conclusion

Changes in the broadleaved woodlands of Cumbria over the last 50 years have, on balance, not been significant: conversion to coniferous species has not occurred on a large scale. The afforestation of open land with conifers, large areas of which are now reaching harvestable age, is contributing to Britain's timber industry.

Within Cumbrian woodlands, there are still many unresolved problems of both scientific and practical importance.

References

Brown, A.H.F., Pearce, N.J. & Robertson, S.M.C. 1979. *Management effects in lowland coppice woods: vegetation changes resulting from altered management in ancient coppice woods in the west Midlands.* (Unpublished report to Nature Conservancy Council.) Grange-over-Sands: Institute of Terrestrial Ecology.

Bunce, R.G.H. 1982. *A field key for classifying British woodland vegetation, Part I.* Cambridge: Institute of Terrestrial Ecology.

Bunce, R.G.H. 1989. *A field key for classifying British woodland vegetation, Part II.* London: HMSO.

Forestry Commission. 1984. *Census of woodlands and trees, County of Cumbria.* Edinburgh: Forestry Commission.

Halliday, G. 1978. *Flowering plants and ferns of Cumbria.* (Centre for North-West Studies occasional paper no. 4.) Lancaster: University of Lancaster.

Hill, M.O. 1979. The development of a flora in even-aged plantations. In: *The ecology of even-aged plantations*, edited by E.D. Ford, D.C. Malcolm & J. Atterson, 175–192. Cambridge: Institute of Terrestrial Ecology.

Lake District Special Planning Board. 1978. *The broadleaved woodlands of the Lake District.* Kendal: LDSPB.

Tansley, A.G. 1949. *The British islands and their vegetation.* Cambridge: Cambridge University Press.

Whitbread, A. 1985. *Cumbria inventory of ancient woodlands.* (Provisional.) Peterborough: Nature Conservancy Council.

Yapp, W.B. 1953. The high-level woods of the English Lake District. *North western Naturalist*, **24**, 188–207 and 370–383.

The management of woodlands in Cumbria

J C Voysey
Forestry Commission, Grizedale

4.1 Introduction

It is a truism that foresters have to manage what their forebears planted. Satchell (see page 2) has set the scene for many of the woodlands which we have to manage in Cumbria today, and to these must be added the plantings of landowners over the last 150 years. However, today's objectives may be very different from yesterday's.

4.2 The legacy

Fourteen-year-old coppice oak (*Quercus* spp.) managed for bark and charcoal, yielding £38 ha^{-1} in 1808, has become today's stored high forest worth £2,250 ha^{-1}. However, now it is probably priceless for its wildlife and landscape value. What was planted for one end product may be reaped for a very different market.

Ship-building oak had become scarce by the mid-17th century. Such trees should have come from the coppice-with-standards woods, but values of underwood suggest that there was little incentive to owners to recruit new standards, or 'spires' as they were called. A few giants still grace our landscape and that is their value.

Between 1782 and 1784, Agnes Ford of Grizedale is recorded as having planted over 77 000 trees, mainly oak and ash (*Fraxinus excelsior*). She married Henry Ainslie, and their son Montague planted a further 1.25 million European larch (*Larix decidua*). The descendants of those crops are still to be seen and are managed as part of the broadleaved working circle. Sycamore (*Acer pseudoplatanus*) was also planted, no doubt for the bobbin mills; in 1882, seedlings cost £2.50 per thousand.

European larch planted to produce pitprops for the iron ore mines in the early part of the last century never saw those cavernous depths, the majority seeing service in the two world wars in a beleaguered timber economy. One plantation, escaping by the lottery of its birthday, is now a seed stand. This old Lakeland larch came from a good stable; it looks like Atholl seed, handed on from one landowner to another. The remainder, handsome trees though they are, are too knotty for the boatskin market.

War-time reserves of timber were once this country's prime object of management; today, the object is to save foreign exchange and ensure supplies. Fortunately, the silviculture for both objectives is the same. From the time that 'deals' began to be imported in quantity from the Baltic ports, Cumbrian landowners found it commercially worthwhile to grow them for themselves, and 19th century account books show the price of transplants of various conifers. Landowners predicted the conifer economy, and planned accordingly. Our timber industry, once almost 100% broadleaved, became 90% coniferous over a relatively short period and has guided forest management in Cumbria as much as elsewhere in Britain.

The 'new' forests of Thornthwaite, Ennerdale and Grizedale (though there is nothing new about the latter) arose because of concern for unemployment on the west coast. Let us not forget that Rider Haggard in the 19th century urged the formation of a state forest service in this country for that express purpose. In hill land this objective can still be relevant, but in forestry now, as in every other industry, efficiency is a synonym for job losses.

4.3 Management today

Cumbria, the second largest county in England, today has a little under the national average of woodland. Bunce (see page 12) has described the tree species composition and age class structure. Cumbrian woods are more than 2:1 coniferous/broadleaved (Forestry Commission 1984). Sixty per cent of Cumbrian woodland is privately owned and the remainder is managed by the Forestry Commission (FC). Of the privately owned woodland, about a third is managed under grant-aided management schemes from the FC.

Soil type is an extremely important variable which has a strong influence on woodland management. The different types of woodland in Cumbria can be related to the soil associations of the Soil Survey of England and Wales (Jarvis *et al*. 1984). The oak woods are found mainly on the upland brown soils of the Manod association in the Lake District, the gley soils of the Clifton association and the stony soils of the Powys association. Sitka spruce (*Picea sitchensis*) is the main species grown on those former, least economic, sheep farms of the Bewcastle Fells, on the gley soils of the Wilcocks association. The application of phosphate one and seven years after planting speeds the establishment of this species. Scots pine (*Pinus sylvestris*) is a very versatile species. It grows well on a large number of sites in Cumbria and has been used to nurse young broadleaved crops. The best crops are found in the drier eastern half of the county on brown sands of the Bridgnorth and Newport associations and podzols of the Crannymoor association. The larches are less tolerant of exposure and wet soils, but have nevertheless been planted widely with good effect.

The markets, both for sawlogs of all species and for small roundwood, especially coniferous, still exist today. The main market for small hardwood remains firewood. Two problems faced by tree growers in Cumbria, as in much of north-west Britain, are wind and deer. Good silviculture will help alleviate the former, while heavy culling and even fencing are required to prevent the large roe deer (*Capreolus capreolus*) population, and in places red deer (*Cervus elaphus*) herds, from causing unacceptable damage to young trees, especially broadleaves. Sheep trespass and damage can be a problem locally, but is a social issue.

Silviculture is still aimed at producing sawlogs for 'deals', as it was nearly 200 years ago. Some foresters may try to produce them in 30-year-old trees, and who can blame them for trying; farmers try to do the same with their products. The zest for quick returns, however, is likely to rebound on the grower, whether it be tasteless young beef or knotty cores of juvenile wood and an unacceptably low yield of M75 stress-graded timber. Pioneers take risks. Quality always sells.

Over 25 years ago, as a result of the 1960 World Forestry Congress in Seattle, the FC began developing multiple land use at Grizedale. Land is managed for the production of food, timber, wildlife and recreation, and the forest is designed accordingly.

4.4 The future

Extensive coppicing has now probably gone forever, a landscape blessing in disguise. Low-input coppice systems, to produce fuel for the domestic hearth of the owner of the small woodlot, have a place for those prepared to learn the art. Ship oak may be required no longer, but fat clean butts have always sawn and sold well, and, as such, oak, beech (*Fagus sylvatica*), ash, sycamore and cherry (*Prunus avium*) should continue to do so. Small-leaved lime (*Tilia cordata*) and wych elm (*Ulmus glabra*) are minority timbers in our woods, which is all the more reason for them to be of quality. Sawmilling requires the least energy input for its product, compared with reconstituted wood in its various forms, whether it be chips or laminates glued together again. This is a major advantage for the quality log.

Present markets suggest that high-forest systems producing coniferous and broadleaved sawlogs, according to site quality, will continue, with clearcutting systems predominating. Large deer populations and wind make shelterwood systems difficult to manage.

Different owners will have different objectives or different priorities for objectives. The National Trust, which owns 10% of the Lake District's woodland, is likely to manage for long-term landscape conservation, while the North West Water Authority, with 4% of the Lake District's woodland, has to protect its catchments from soil erosion by, amongst other things, long rotations. Both organizations need to produce commercial timber in order to fund their objectives. The proportion of broadleaved woodland is likely to continue to rise, in line with the multiple land use policies practised by the FC at Grizedale.

The upland peats and exposed escarpments of the Pennines and the high elevations of the Lake District with their thin soils are generally unsuited for commercial tree growing and could account for some 20–25% of the county. Elsewhere, forestry could be an alternative land use. However, it is unlikely that land surplus to the growing of cereals will be available for broadleaved forest, contrary to the hopes of the Chairman of the Nature Conservancy Council (Nature Conservancy Council 1985). The arrival of the grey squirrel (*Sciurus carolinensis*) is going to make this even less likely. Hill land is much more likely to go out of agriculture and to find a ready buyer in the forester. Even if hill land were to be abandoned and revert to scrub, something that is feared by some, it would be no bad thing because it would then eventually revert to woodland. The Lake District is full of eroded sites from which all sheep should be excluded, and there are plenty of examples of woodland reversion where this has occurred. One example is Launchy Gill, above Thirlmere, which is now a Site of Special Scientific Interest. Such 'management' has, in the past, been fortuitous, but it could become a positive objective.

4.5 Conclusion

What is this strange attitude to commercial forestry? Woodlands have always been commercial enterprises and, when they ceased being commercial, they disappeared. From the time we sank to an all-time low of 4% woodland in Britain and rushed to live in towns, we have been silviculturally illiterate, and our foresters have been vilified. It may be necessary for privately owned game woods to remain undisturbed, but, thanks to our state forests, the public is at last beginning to value its woodlands, its access to them and the opportunity to observe the wildlife they contain. These values have led to new objectives of management and to important changes in the way we manipulate tree crops; altering them to increase the numbers of wildlife species has improved their appearance, and *vice versa*.

Landscape architects are brought into management teams, the FC has a £4.6 million conservation and amenity subsidy, and glossy propaganda on conservation is published by the FC, Timber Growers United Kingdom, Economic Forestry Group, and others.

The Wildlife and Countryside (Amendment) Act of 1985 amends the Forestry Act so that the FC shall endeavour to achieve a reasonable balance between timber production, the enhancement of natural beauty and the conservation of wildlife. Legislation has, again, caught up with the management some foresters have been practising for a long time.

References

Forestry Commission. 1984. *Census of woodlands and trees, county of Cumbria.* Edinburgh: Forestry Commission.

Jarvis, R.A., Bendelow, V.C., Bradley, R.I., Carroll, D.M., Furness, R.B., Kilgour, I.N.L. & King, S.J. 1984. *Soils and their use in northern England,* (Bulletin 10.) Harpenden: Soil Survey of England and Wales.

Nature Conservancy Council. 1985. *Eleventh report.* Peterborough: Nature Conservancy Council.

Harvesting management options

D E Davis
Cumbria College of Agriculture and Forestry, Penrith

5.1 Introduction

Timber harvesting is an expensive and dangerous activity. A growing tree is of a most inconvenient shape and size when one is considering transferring it from its vertical position half-way up a Cumbrian fell to the horizontal in a sawmill or pulpmill, maybe several hundred miles away. These problems are compounded by the terrain, climate and weather, to the point where statistics inform us that forest harvesting is the most dangerous industry in the UK.

Cumbrian woodlands have no unique features for the harvester, but the high rainfall and frequency of steep slopes place them amongst the more difficult woodlands in Britain.

I will be restricting this paper to plantation-grown conifers and broadleaves and excluding mature hardwoods, whose value and size make them a special case. I am also assuming that anyone harvesting wood will be attempting to show some financial return, or at least to minimize expenditure.

5.2 Thinning

Any given area of adequately stocked woodland will produce much the same volume of harvestable timber, whether it is thinned regularly and then felled at an economic rotation age, or left unthinned until clear-felled. However, the financial advantages of thinning should be:
i. some return much earlier and at regular intervals during the rotation;
ii. the higher value of the larger stems which will comprise the final crop.

There are, of course, many other aspects to be taken into consideration when making the decision of whether or not to thin, including species, soil, growth rate, stability, road costs and harvesting costs. All these aspects can be defined or calculated, but the possible landscape and environmental factors are less easy to quantify. It is desirable, if not essential, that the thin/no-thin decision be taken before the crop is planted so that the correct management options can be implemented throughout its life.

The marketable wood obtained from a thinning operation is more expensive to procure and will probably sell for less than the produce from a clearfell. It is, therefore, essential that the best market is identified before

the operation is planned, let alone started. There is also the danger of wasting time and money by felling material which cannot be sold.

An adequate volume of saleable timber will have to be produced, which in the present economic climate would mean at least 40 m^3 ha^{-1}, for the operation to stand a chance of being profitable. However, it is also essential that a sufficient stock of vigorous stems is retained to ensure future production.

A simple systematic or pattern thinning may suit the crop and the forest manager's objectives. Such thinning will always be cheaper, but the cut trees will include a number which would have been best left for the final crop. Some of the more esoteric thinning systems make the cutter's job more difficult, and therefore increase harvesting costs.

Extraction routes should be planned and marked, bearing in mind the capabilities and requirements of the machine that will be using them, before the trees are marked for thinning. However, with adequate supervision, it may not be necessary to mark the trees for thinning; the interest and skill of the cutter are often underestimated.

5.3 Felling

Compared with thinning, felling produces larger volumes of timber, which are obtained more rapidly and which sell for higher prices. However, timing is of the essence because growth rate slows dramatically with age. Thus, the economic rotation age of a tree crop is very much shorter than the natural lifespan of the species, and financial criteria demand its replacement to maintain increment production.

Fluctuations in market prices are also critical. One benefit of the long rotation of tree crops is the facility to advance or retard final felling by several years in order to take advantage of favourable timber prices. This single decision can have more influence on profitability than any other taken in the lifetime of the crop.

Large-scale felling operations can normally be made more efficient, and therefore more profitable, than small patches, shelterwood, or true selection forest systems. There are great economies to be made in arranging an annual felling programme in one area, although the effect upon the landscape may be quite dramatic. The economic disadvantage of the small-

5.7 Conclusion

Whilst woods are many different things to many different people, they are primarily a renewable resource of cellulose which requires expertise, organization and considerable expenditure to harvest successfully.

Reference

Helliwell, D.R. 1984. *The economics of woodland management.* Chichester: Packard.

Woodland produce – markets and marketing

A S C Meikle
Economic Forestry Group plc, Kendal

6.1 Introduction

Britain currently imports 90% of its requirements of timber and timber products at a cost in excess of £3,500 million. This import bill is only exceeded by our imported food requirements. Of the 10% of our annual timber requirement produced in the UK (approximately 4 Mm^3), about two-thirds is produced by state forests administered by the Forestry Commission (FC). The remainder is provided by the commonly named 'private sector', which encompasses all sources outside the FC and includes such bodies as local authorities, water authorities and trusts, together with private timber growers.

This overall situation has important implications for the growing and marketing of home-produced timber. When the burden on the balance of payments of such heavy imports is taken into account, together with overproduction from British agriculture, a move to lessen our reliance on imported timber becomes attractive. It must also be considered that Britain has great potential, in both land and expertise, to produce greater timber volume. Britain currently has 9% of its land area under forestry, compared with an average of 22% for countries in the European Community (EC). It is interesting to note that land under agriculture in Britain is in excess of the EC countries' average by about 16%.

The most important influence that the high import situation has on the marketing of home-grown timber is on prices. With a market dominated by imported timber, the prices paid for UK timber are largely governed by imported timber values. The availability of markets to UK growers is largely dictated by the availability of overseas supplies. Overseas supplies can be affected in many ways, including the rates of foreign currency exchange, and these can have a significant effect on the home marketing situation.

6.2 UK private sector marketing

In this paper, I am principally concerned with that share of the home market occupied by the private sector grower. In volume terms, the private sector currently produces about 1.2 Mm^3, and it is interesting to note that the production of broadleaved timber in the UK originates almost entirely from the private sector. With such a small share of the total UK market, it is fair to assume that the private grower has to be especially astute and professional in his timber sales, if he is to remain afloat and produce a reasonable return from his woodland investment. Regrettably, this is quite simply not the case for a large section of the private sector. There is very little common approach to marketing by growers, with the result that the market is fragmented and haphazard, and thus there is little or no return on the original investment. This situation is not only unsatisfactory to the grower but also to the market customers, who find it difficult to operate in a climate where raw material supplies are erratic and far from guaranteed, and results in lower prices being paid for timber. There is also an alarming reliance by private growers on their customers (especially sawmills) for guidance and direction in the marketing of timber. It is obvious that buyers will exploit such a weakness on a seller's part, with further diminished returns to the grower and an undermining of the total private sector marketing situation. Regrettably, there is no 'Timber Marketing Board' or similar national marketing body operating in the UK forestry industry, although one has been proposed on several occasions.

This undesirable situation has been recognized for many years and has been the subject of several reviews and reports, such as the Watson Committee report of 1956 and an Economic Investigation Unit report in 1970. There is still a strong desire by many private growers to maintain their complete independence, thereby sustaining the free market situation: unfortunately, this is not economically to their advantage. A better degree of co-operative marketing is offered by the larger management companies, and there is no doubt that woodland owners who market their timber through these organizations enjoy greater security and much improved returns for their timber, by taking advantage of the pooled resources, bargaining power and professional expertise.

The interests of private growers are represented nationally by Timber Growers United Kingdom. This organization has the potential, with increased membership, to help resolve some of our current marketing problems. However, before this can happen, there will need to be far greater common purpose and co-operation demonstrated by its members.

It has already been mentioned that the private sector provides almost the whole of the UK broadleaved timber output. This amounts to over 40% of our annual hardwood requirement and, in percentage terms, our self-sufficiency in this respect is increasing. However, the increase is caused almost entirely by a falling off of overall annual consumption. The furniture industry, for example, is taking increasing advantage of the availability of cheaper softwood panel products, such as laminated chipboard, in place of traditional solid section hardwood.

However, there will always be a demand for good-quality broadleaved timber. Sound management and marketing techniques are especially important in encouraging improved economic returns from broadleaves grown on suitable lowland sites. These improved returns will, in turn, improve investment in broadleaves and lead to an increase of the overall area planted in the UK. Not only will this secure an important raw material asset but it will also represent an important environmental gain.

Sound marketing techniques are equally as important for 'environmental' woodlands as they are for commercial woodlands. There is evidence that some foresters responsible for the management of such woodlands place little emphasis on the effective marketing of timber. This is to be regretted as it leads to the needless loss of revenue that could be usefully employed in helping to extend their main sources of income.

6.3   Principal marketing outlets

The markets for timber tend to fall into three main groups.

i. **National** Typical markets under this heading are the major pulpmills, producing paper, and the panel mills, producing processed timber panels such as chipboard. Examples in the north of England are the Thames Board Mills at Workington and Egger UK at Hexham.

ii. **Regional** Under this general grouping are found the larger sawmills (producing sawn products such as construction timber and pallet boards), turned wood manufacturers and the larger fencing manufacturers. A typical example in Cumbria is Lowther and Croasdale's pallet mill at Penrith.

iii. **Local** There are many outlets in this group including the smaller sawmills, fencing manufacturers, rustic craftsmen, chockwood mills and firewood merchants.

Markets in the local category are normally directly accessible to the smaller timber grower. However, markets in the regional and especially national categories often have large and sustained supply requirements, making their direct supply beyond the scope of the smaller grower. Timber finds its way to these out-

lets *via* a timber merchant who is in a position to consolidate supplies of timber from smaller growers and to negotiate a level of supply and price with the customer.

6.4   Marketing

The marketing of timber is an integral part of the management of a woodland; it must be planned within the overall framework of the woodland and not carried out on an *ad hoc* basis, as is so often the case. Planning normally takes the form of production forecasts, the construction of which requires a sound forestry knowledge. By comparison with other industries, production planning in forestry is often long term and may involve forecasting yields of timber 20 years hence. It is very important to have a sound knowledge of markets and timber values in order to produce a production plan that will ensure the best silvicultural and economic return from the woodland.

6.5   Points of sale

There are three principal points.

i. **Standing** In this case the trees are simply sold where they stand, it being the purchaser's responsibility to fell and remove the trees. This is the 'safest' method of sale for a vendor, as he passes on to the purchaser all the risks associated with harvesting and marketing. The vendor also passes on the risk of timber quality, as this does not often become apparent until the trees are felled. However, by factoring most of the risks and costs to the purchaser, the returns are correspondingly lower.

ii. **Roadside** If the grower has the means of felling and extracting timber to the roadside, he may wish to sell the timber at that point. Obviously a lot of the risk and cost is now transferred to the grower, and he can therefore expect to receive a higher price for his timber from the purchaser. It must also be remembered that, if the timber is of poor quality, this will be more easily seen at the roadside, with a corresponding fall in value. Thus, if timber quality is suspect, it is often safer to accept a smaller return by selling standing.

iii. **Delivered to market** Having felled and extracted the timber, the grower may elect to carry out a complete marketing operation and to sell the timber delivered to his various customers. To do so, he will need a very thorough knowledge of harvesting and marketing, together with the necessary resources, and he will be carrying the entire risk and cost of the operation on his own shoulders.

Davis (see page 22) describes these three options from a harvesting viewpoint.

6.6 Methods of sale

There are three principal methods of sale.

i. **Negotiation** This is commonly used in the private sector and is a regular cause of problems. Agreement is simply reached between the vendor and purchaser as to the price and point of sale. All too often such agreements are verbal only and arrived at without proper knowledge, resulting in a dispute from which the vendor rarely emerges unscathed. Negotiation is perfectly valid, providing it is carried out between parties who are knowledgeable and who understand what is being sold. Any negotiation sale must be backed up by a written agreement, in the form of a contract of sale.

ii. **Sealed tender** This method is often used where timber is sold standing or at the roadside. Details of what is being sold and the conditions that will apply are circulated to interested parties, who then submit written offers. This method has the advantage of introducing competition between prospective purchasers, and prices are therefore more likely to reflect the true market value of the parcel of timber being sold.

iii. **Auction** The sale of private sector timber by auction is relatively uncommon. The FC, on the other hand, markets a large part of its annual production at regular auctions and has offered private sector growers the opportunity of selling timber at its auctions. Auction sales expenses are inclined to be high, making it uneconomic as a means of selling smaller parcels of timber individually. However, auction sales do attract considerable interest, and prices obtained can be high when timber is in demand. The fragmented nature of the private sector is much to blame for the failure of private growers to take advantage of this method of sale.

As has already been mentioned, it is vital that any sale is properly supported by a legally binding and signed contract between the two parties involved. The contract must, at the very minimum, clearly identify what is being sold, the price, when it should be paid, and when work is to start and finish. It should also provide details of access to the work site, any particular do's and don'ts and the penalties and methods of arbitration that might be applied. It is also important to remind the purchaser of his obligations under the Health and Safety at Work Act.

6.7 Conclusion

Because of its small size, the home-grown timber market is dominated by the availability and cost of imported timber and timber products. Private sector marketing is especially fragmented, with the result that financial returns on timber are often poor and not representative of the long-term investment in the crop. The current situation could be improved by more co-operation between woodland owners, but a greater step forward will be achieved by better education and understanding of production planning, markets and marketing principles.

The economics of woodland management in Cumbria

R E Shapland
North West Water, Thirlmere

7.1 Introduction

The basic objective of woodland management is one of sustained yield. This yield may be of timber products or benefits, such as landscape, recreation or nature conservation. Unlike other European countries, the UK has never been prepared to commit the resources needed to maintain its woodlands in a healthy and productive state; it has always expected the woodlands themselves to provide a substantial contribution to their costs. It is possible, without doing calculations, to see which woodlands are less successful economically by identifying the long-term national trends in their structure and quality. Some broadleaved woodlands have been felled and replaced with conifers, while many of those remaining have lost their protective fences and are now damaged and prevented from regenerating by grazing farm animals. Little thinning has been undertaken resulting in a loss of timber quality, and trees are often allowed to die without being replaced. Coppice is rarely worked and hedgerow trees continually disappear. New planting has tended to be on the poorer land with coniferous rather than broadleaved species.

There are two sides to the economics of woodland management, expenditure and income, and it is their relationship which is of vital importance to the future of our woodlands in Cumbria. Both factors vary for a variety of reasons, some of which I will now discuss in greater detail.

7.2 Size and shape of woodlands

Fencing for woodland protection against farm animals, rabbits and deer can be the largest single item of expenditure over the lifetime of the forest. Table 1 demonstrates that, as the size of the woodland falls, so the cost of fencing per hectare increases, and this has no effect whatsoever on the income per hectare from the timber produced.

Table 1. Cost of erecting stock fencing around square woods of different areas, assuming a cost of £1.70 m^{-1}

Area (ha)	Length of fence (m)	Total cost (£)	Cost ha^{-1} (£)
100	4000	6800	68
25	2000	3400	136
10	1260	2140	214
5	890	1510	302
2	570	970	485
1	400	680	680

The figures in Table 1 relate to square woodlands, but costs usually rise for a woodland of a different shape. For example, a rectangle of five hectares, 100 m by 500 m, would have a fence line of 1200 m, and the cost of fencing against farm animals would be £408 per hectare (an increase of £106 ha^{-1}). Where there are small woodlands with long irregular fences, the costs increase along with the complexity of the woodland outline. The costs quoted relate to easily erected fences designed to exclude farm stock, but costs can rise by 50% where the fences are in remote or difficult situations, because of the problems of transporting men and materials. Fences designed to exclude rabbits and/or deer will also increase costs to £5 per metre, or even more in difficult areas, with costs per hectare rising to as much as £3,750.

Woodlands within Cumbria suffer from exposure because of their altitude and proximity to the sea. Smaller woodlands have a greater proportion of edge trees which can be affected by exposure, thus reducing growth rate and timber yield. Curvature of stems and the heavy rough branches always found on the boundaries of exposed plantations reduce the quality of the timber.

The cost of timber harvesting is also affected by the size of the woodlands, as the time taken to move equipment between sites is still the same, irrespective of the size of the wood. Difficulties arise when the timber has to be distributed to several markets, and less than full loads are produced at each location. Time, and consequently money, is lost in collecting from several sites.

The cost of regularly inspecting the woodlands for protection and management purposes is often overlooked, and insufficient time may be set aside for this purpose, resulting in damage and financial loss. It is obvious that it takes less time to inspect one 100 ha wood than 100 woods of one hectare scattered over a large area.

Where groups of woodlands are managed as a unit, as was the case with country estates in the past, the difficulties were not so great because the larger, more economic woods supported the less economic. Perhaps more significantly, the sporting interest of the landed gentry ensured that funds were available for maintenance. Changing attitudes to blood sports and increasing numbers of visitors have reduced the money available from this source in Cumbria.

7.3   Species of trees

The costs of purchasing different species for planting vary, with broadleaves generally costing more than conifers by at least 30% and as much as 100%. This extra cost can be compounded by greater post-planting losses of broadleaved species in the more inhospitable parts of Cumbria.

Timber volume production also varies with species, and this factor has a considerable influence on income. Sitka spruce (*Picea sitchensis*) yield is perhaps ten times that of oak (*Quercus* spp.).

7.4   Site quality

Cumbria has a wide range of site types, from fertile lowlands to poor rocky highlands. The former can have high weeding costs which are compensated for by higher production of good-quality timber; the latter may suffer from tree deaths and poor timber production, which may only be fit for firewood.

The replanting of land previously under forest, but cleared and used for agriculture for generations, is often difficult due to the destruction and loss of soil and nutrients by erosion and sheep grazing. It may take many, perhaps hundreds, of years to recreate the type of soil suitable for the growth of good-quality, healthy forests.

7.5   Accessibility

We all appreciate the rugged and varied terrain of the Lake District. However, this terrain can have a serious adverse effect on the costs of management, maintenance and timber harvesting, and thus reduce the possibility of the woodlands being economically viable. A shorter working week, coupled with a requirement for better and safer working conditions, has increased the need to provide transport to the work site. Such provision necessitates road access in areas where terrain can create enormous problems for road construction, as well as landscape design.

It may prove to be too expensive to manage some woodlands in the future, unless some form of vehicle access is provided, as they will not be visited regularly and fence breakage will go undetected for many weeks. As a result, entry will be afforded to farm animals, and the woodlands may be totally destroyed in times of bad weather.

7.6   Distance from markets

'Local' markets tend to produce what are termed traditional products, such as hurdles, charcoal, bobbins. Demand for these products has not only declined within Cumbria, but nationally.

Wood-using plants have become larger and are usually sited near centres of population, which leaves Cumbria at a disadvantage, especially with broadleaved timbers which are not abundant north of the River Mersey. Additional haulage costs to far-away markets reduce the net income of the woodlands. The viability of the woodlands within the county is likely to rely on the expansion of local industries within Cumbria itself, so it will be important for local authority planning departments to take this factor into account when formulating their criteria for planning permission. A finished product usually takes up less vehicle space than raw unprocessed timber, as timber is invariably wasted in any manufacturing process. Costs of transport are therefore reduced, there is less traffic and increased income is brought into the county, all of which can benefit the locality as well as the woodlands themselves.

7.7   Conclusion

In this paper, I have endeavoured to show that there is no single answer to the problem of woodland economics. Whether or not a woodland is capable of producing enough income from its timber to cover its expenditure depends on many factors, some of which I have mentioned. The economics of each woodland must be calculated individually, remembering that circumstances are bound to change over the lifetime of the trees.

We may wish to ignore economics, especially where woodlands are supposed to exist for landscape, recreation or nature conservation purposes, but to do so reduces our ability to manage and maintain the greatest possible area of woodlands.

The silvicultural management of woodlands for timber production is basically the same as management for other purposes, and involves the same forest operations which need to be carried out efficiently and economically. It follows that, if income is to be foregone by choosing species of trees and silvicultural systems which do not give the highest financial return for the site, the management must be of the highest quality to achieve the most efficient use of resources. Otherwise, our woodlands will continue to decline.

Finance for woodland maintenance always appears to be limited, so some account must be taken of economic factors when deciding priorities. In general, the larger lowland woods on good sites, capable of growing high-quality timber, with good access and using simple silvicultural systems, will be most able to contribute to their costs. At the opposite extreme, small woods, of irregular shape, on poor exposed sites inaccessible to vehicles and managed on complicated systems, will either fail or require large inputs of money for a limited result.

The long-term future of the Cumbrian woodlands depends on good and efficient management, backed up by a vigorous local timber industry.

The social implications for rural populations of new forest planting

M Whitby

Department of Agricultural Economics, University of Newcastle-upon-Tyne

8.1 Introduction

The current interest in rural employment stems from the relative profitability of agriculture and forestry; for financial reasons, there is substantial enthusiasm for extending the area of forest in this country. It is not my purpose to deal with this aspect of the case for forestry, but I should acknowledge that arguments for extending our area of forest seem to be stronger now than they have been for some time. This strength of the forestry case rests, to some extent, on current doubts about the future of agriculture. It does seem that we may be at a turning point in the implementation of the European Community's Common Agricultural Policy (CAP), which implies that the financial returns from most agricultural activities are likely to deteriorate over the next few years. If this decline occurs on hill farming land, then the rate of conversion to forestry may accelerate. There is now considerable policy interest in converting land from agriculture to forestry as a means of reducing agricultural surpluses (Commission of the European Communities 1985, 1986).

Apart from the rather broad issues relating to the relative efficiency of agriculture and forestry, there is an important strand in the afforestation argument which asserts that afforestation also increases rural employment. The supporting evidence usually quoted comes from dividing present *levels* of forest employment by the forest area. Such comparisons yield broad estimates of the density of forest employment of one man-equivalent per 100 hectares in England and one per 111.5 ha in Great Britain. Such estimates are compared with the (often lower) employment levels in hill farming, and the conclusion is drawn that forestry is a 'better' employer than agriculture. However, I must insist that, especially in the case of new forest planting, such a procedure is completely inappropriate and can only mislead.

To compare the effects of new forestry with those of agriculture as a competing land use, it is necessary to:

- project the labour requirements of forest operations throughout a complete rotation
- project the labour use in agriculture throughout the same period.

This paper follows these steps for Cumbria, drawing on a detailed study undertaken for the Countryside Commission (Laxton & Whitby 1986) which examined the prospects for employment in forestry in the three northern counties of England.

8.2 Labour requirements: present and future

Agriculture

Direct employment in agriculture can be assessed by projecting past trends. As is well known, the direction of total employment in agriculture has been downwards for a long time, but what is important here is the trend of employment in hill farming. In order to measure the trend of employment in agriculture, we examined the parish summaries of the agricultural census over the period 1960–83. Data were available for six different years in that period, and summaries were available for some 60 parishes in Cumbria. The number of regular hired workers per thousand hectares was regressed on a linear time trend to estimate the average rate of decline over that period.

That rate of decline was then averaged together with the number of farmers, as revealed in the recent censuses, to produce a long-run rate of decline. In Cumbria, the average number of regular workers and farmers per thousand hectares in 1983 was found to be 14 and the rate of decline 1.4% per annum.

Forestry

For forestry, current levels of employment reveal little with regard to the future, because existing plantations in Cumbria have not yet reached an evenly mixed age structure. Accordingly, it is necessary to consider the requirements of new planting in terms of the labour used for the various operations throughout the forest rotation. Such an approach has been developed by Inglis (1978), and his method has been used by others examining the forestry sector (eg Centre for Agricultural Strategy 1980). The essence of the Inglis technique is to list the activities, year by year, associated with a particular management regime and to estimate their requirement of labour. These may then be grossed up to actual scheme requirements in each year. In his original work, Inglis considered a block of 2900 hectares which was developed over 50 years, with equal amounts of land (58 ha) being planted each year. The result was that his scheme required the length of a full forest rotation before it reached maximum employment. In our study, we applied work rates to schemes of different sizes, but here I will concentrate on a scheme of 500 hectares, planting Sitka spruce (*Picea sitchensis*) of Yield Class 12.

A crucial part of the Inglis approach is the accurate specification of the forest management regime and the precise estimation of the labour requirements for each operation. Obviously, these two attributes are not fixed at the time of planting, but it is possible to state what activities are currently undertaken on plantations of different ages and how much labour they use. We were fortunate in being able to obtain data on these two aspects from Forestry Commission District Managers, and those data formed the basis of our estimates. Clearly, such current labour requirements are subject to growth in productivity, but that question will be addressed later.

In our discussions with forest managers, it became clear that a major variable to be considered in forest management regimes was whether or not thinning was to be practised. Current Forestry Commission practice is to omit thinning in most cases, but, if thinning is included, then the total employment generated by planting is substantially increased; moreover, it is brought forward in the rotation. Figure 1 compares two different management regimes on a 500 ha plantation over 60 years. The annual requirement of labour under both regimes for the first ten years peaks at six man-years. Thereafter, without thinning, there is hardly any employment until year 40. For the five years of clearfelling, 40–45 men are employed per year. By contrast, there is a substantial amount of employment with thinning (seven to eight man-years) from years 19 through to 39 in the rotation, after which there is a gap followed by the peak of employment (26–30 man-years) in years 44–49.

For these two regimes, we may thus tabulate the number of man-years per decade. Table 1 indicates

Table 1.   Labour requirements for 500 ha of new forest planting per decade

| Years | Man-years | |
	With thinning	Without thinning
1–10	37.4	37.4
11–20	9.2	1.2
21–30	83.6	1.2
31–40	73.0	41.3
41–50	173.1	226.1
51–60	10.2	0
Total	386.5	307.2

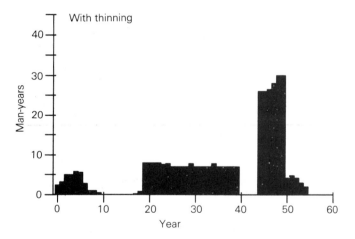

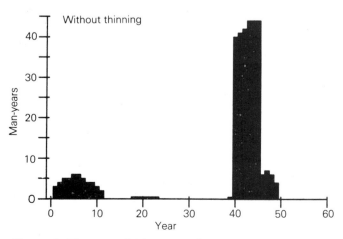

Figure 1. The annual labour requirements throughout the rotation of a 500 ha plantation, with and without thinning.

that, although the labour requirement of planting is moderate throughout the decade of planting, thereafter there is virtually no employment if thinning is not practised, until the felling stage is reached in the fifth decade of the rotation. Where thinning is practised, the labour requirement is high in the third and fourth decade, and then peaks in the fifth decade. Comparison of the totals shows that thinning thus requires some 25% more labour over the whole forest rotation than a non-thinning regime; furthermore, it substantially brings forward the amount of employment generated by afforestation.

Indirect employment

It is customary for advocates of afforestation to emphasize the indirect employment generated in timber processing, which could be an important contribution to rural employment. However, there are two major qualifications which must be added to such an argument. First, the employment generated as a result of forest planting is mainly associated with processing and, therefore, will not appear until some timber is produced. This production would be in year 20, where thinning is practised, but in the absence of thinning it would not occur until year 40. Second, the indirect employment generated in forestry does not have to be located in rural areas. Indeed, much of it is likely to be urban. In its assessment of the employment in timber processing in the late 1970s, the Forestry Commission (1978) estimated that only 25% of the jobs generated in processing timber would be located rurally.

Similar arguments can be deployed with regard to indirect agricultural employment, although one minor difference is worth noting, relating to the production of store stock on the hills. Insofar as the livestock produced are unfinished, that is in need of further fattening before slaughter, then it is evident that their further processing in agriculture will generate further *rural* employment. Such employment will not necessarily be in the same area as that in which the sheep were born, but it is nevertheless rural and may be significant. A partial offset to this argument would be the current tendency for store rearing farms to fatten their own livestock as far as possible. The extent of this practice is not widely documented, but the farm management survey data collected at Newcastle show that 28% of lambs sold from hill farms were in fat condition and the rest were sold for breeding or for fattening elsewhere (Whitby 1986). Otherwise, similar arguments apply to those in the forestry context, ie processing of animals for human consumption is mostly undertaken in substantial plants, the majority of which are urban-based.

It is tempting to conclude that the indirect employment effects of agriculture and forestry might be of similar importance to rural areas. What is absolutely unknown is the way in which this importance might change, particularly for the products of forests as yet unplanted. Accordingly, this aspect of employment will not be discussed further.

8.3 The net effect of conversion from agriculture to forestry

If, for convenience, we are to consider forestry and agriculture as mutually exclusive uses of land, then it is evident that forest planting must displace agricultural employment. Therefore, the net effect would be the total forestry employment minus the lost agricultural employment.

Because the direction of conversion is from agriculture

to forestry, and the scheme that we are considering takes five years to implement, the agricultural employment loss is phased over that period. The lost agricultural employment will increase to a maximum in the fifth year of the scheme and continue thereafter at the same level. We may thus deduct the full amount of employment lost from agriculture from the full generation of employment by forestry in order to reach conclusions about the impact.

Comparing these amounts of employment over the whole rotation with the estimated agricultural employment, the first row of Table 2 shows that in both cases more agricultural employment will be lost than is gained through forestry. In fact, the difference between forestry with thinning and agriculture is negligible over a whole rotation, although there is a more substantial net loss of employment when thinning is not practised.

Table 2. Projected man-years of employment in 500 ha of upland Cumbria, over 60 years, at various levels of labour productivity

Annual percentage change of labour productivity	Agriculture employment loss due to forest planting	Forestry employment gain	
		With thinning	Without thinning
0	393.8	386.5	307.3
2	221.5	198.1	150.7
4	139.0	112.8	82.7
6	95.3	71.6	52.0
8	70.0	50.2	37.3

The discussion so far has been based on current levels of employment in both agriculture and forestry. Yet we know that in both industries there is a continuous process of downward adjustment in employment as machines become better adapted and as the operations undertaken in both industries require less manpower.

It has already been mentioned that agricultural employment per hectare in upland Cumbria has been declining at 1.4% per annum for the last 20 or so years. Will that rate continue? In addition to the decline in employment caused by the increasing efficiency of labour use, there is also the prospect that hill farms may go out of business. It is likely that CAP will be adjusted and, after the disposal of the milk and cereal surpluses, the next candidates for downward adjustment of price will be beef and sheep. Many commentators believe that a full adjustment of the CAP must also include the

Conservation of flora and fauna

K J Kirby[1] and A M Whitbread[2]

Nature Conservancy Council, [1]Peterborough, [2]Grantham

9.1 Introduction

Much of Britain was once covered by forest, and, while a considerable part of our present flora and fauna does not depend on woodland for its survival, woodland remains probably the richest habitat overall. Roudsea Wood in south Cumbria, for example, holds at least 200 species of fungi, 30 molluscs, 60 spiders and 26 butterflies (P Singleton pers. comm.).

In this paper, we assume that the general case for conserving wildlife is accepted (Nature Conservancy Council 1984), and explore the questions of which species or communities should receive most attention and how their conservation may be achieved. One approach to woodland conservation is to aim to maintain or increase the populations of as many woodland species as possible. However, some species and communities are more abundant and less threatened than others. Hence, a second approach is to concentrate conservation efforts on the rare or threatened species and communities. Lastly, given the predominance of man's influence generally throughout Britain, it can be argued that the more natural communities (where man's interference is less direct or less intensive) should receive priority. These three approaches, emphasizing the criteria of diversity, rarity and naturalness (Ratcliffe 1977), can all be important in different circumstances.

9.2 The diversity of flora and fauna in Cumbrian woodlands

Almost any species may occur in woodland, but some, without necessarily being confined to woods, are more strongly associated with woods or the habitats found within them.

For 'woodland' vascular plants, we may use, as a first approximation, the species listed on the Nature Conservancy Council's (NCC) woodland survey card. There are 404 of these, of which about 360 occur within Cumbria (Halliday 1978), slightly more than occur in Kent which has nearly twice as much ancient woodland. Over 170 of these species have been recorded from the Witherslack Woods in south Cumbria, although 50–70 would be more typical for sessile oak (*Quercus petraea*) woodland or 80–110 for base-rich or poorly drained sites. Even an individual wood, therefore, may contain a significant proportion of the potential woodland flora, albeit generally the common species. The rarer woodland plants are also quite well represented within Cumbria (Table 1).

Table 1. Less common woodland plants recorded in Cumbria (Halliday 1978), excluding possible introductions

Present in only 16–30 ten kilometre squares in Great Britain

Baneberry	*Actaea spicata*
Fingered sedge	*Carex digitata*
Long-leaved helleborine	*Cephalanthera longifolia*
Coral-root orchid	*Corallorhiza trifida*
Dark-red helleborine	*Epipactis atrorubens*
Touch-me-not balsam	*Impatiens noli-tangere*
Angular Solomon's seal	*Polygonatum odoratum*
Downy currant	*Ribes spicatum*
Mountain currant	*Ribes alpinum*
Large-leaved lime	*Tilia platyphyllos*

Present in 31–60 ten kilometre squares

Narrow-leaved bitter-cress	*Cardamine impatiens*
Wood fescue	*Festuca altissima*
Wood barley	*Hordelymus europaeus*
Serrated wintergreen	*Orthilia secunda*
Larger wintergreen	*Pyrola rotundifolia*
Green figwort	*Scrophularia umbrosa*

Present in 61–100 ten kilometre squares

Hair sedge	*Carex capillaris*
Lesser tussock-sedge	*Carex diandra*
Yellow star-of-Bethlehem	*Gagea lutea*
Creeping ladies' tresses	*Goodyera repens*
Limestone polypody	*Gymnocarpium robertianum*
Mountain St John's wort	*Hypericum montanum*
Rock whitebeam	*Sorbus rupicola*
Marsh fern	*Thelypteris palustris*
Bitter vetch	*Vicia orobus*

Based on the number of occurrences (L Farrell pers. comm.) in *Atlas of the British Flora* (Perring & Walters 1982). National rarities (present in less than 16 ten kilometre squares) have been excluded

Cumbria's location and its mixture of upland and lowland habitats mean that it contains both plant species with a predominantly northern distribution in Britain, including wood fescue (*Festuca altissima*), mountain melick (*Melica nutans*), serrated wintergreen (*Orthilia secunda*), bird-cherry (*Prunus padus*) and globe flower (*Trollius europaeus*), and some found mainly in the south, including field maple (*Acer campestre*), bushgrass (*Calamagrostis epigejos*), yellow archangel (*Lamiastrum galeobdolon*), black bryony (*Tamus communis*) and small-leaved lime (*Tilia cordata*). This is also true for some invertebrates, eg the wood ants *Formica rufa* and *F. lugubris*. Woodland-breeding bird populations tend to be less rich than those in southern and eastern England, but the Cumbrian woods are richer in terms of bryophytes and lichens (Ratcliffe 1968). Some of the bryophyte species found in Cumbria are extremely restricted on a European scale.

Sites may be rich in one group of species and poor in others: a coppice wood, for example, may be rich in vascular plants but poor in dead-wood beetles and lichens, because past management has removed all the large trees and dead timber. Dense bryophyte cover tends to occur in woods poor in vascular plants, and some vascular plants such as bilberry (*Vaccinium myrtillus*) and wavy hair-grass (*Deschampsia flexuosa*) may occur only in naturally species-poor communities, especially those of acid soils.

The NCC seeks to protect nationally important woods throughout the country and, in any one county, a range of community types and woodland structures, particularly those which may be closest to the original natural forest cover of the area.

9.3 Woodland types in relation to nature conservation

The large coniferous plantations, such as those of Ennerdale Forest and Thornthwaite Forest, mainly established during this century, are not devoid of wildlife (eg Hill 1979; Moss 1979), and form an extensive new habitat in Britain. The invasion of these sites by woodland species has been monitored (Hill & Jones 1978; Sykes 1981), but it may require several rotations of trees before we can be sure of the full range of wildlife that can thrive in these forests (Nature Conservancy Council 1986).

By contrast, many of the broadleaved woods in the Lake District have been in existence for several hundreds of years at least, and their flora and fauna have largely stabilized to the communities which the present soil, climate and woodland structure will permit. It would be difficult to recreate many of the plant and animal communities that they contain. The NCC therefore places its priority on conserving these communities, rather than recent conifer woods (Peterken 1977; Nature Conservancy Council 1986).

The NCC has most concern for ancient woods, that is sites that appear to have been continuously wooded since AD 1600. There are more species and the species tend to be more abundant in these sites than in recent woodland (Peterken 1983). In particular, the NCC is interested in ancient semi-natural woods where the tree and shrub covers comprise species native to the site and not obviously planted. If planted in the distant past, the several generations of coppice regrowth or self-seeding mean the plantation origin is

no longer clear. The patterns of variation in the present tree and shrub layer and in the ground flora can usually be related as much to environmental factors, such a differences in soils and geology, as to the deliberate, direct effects of man.

In some cases, the ancient semi-natural woods may be direct descendants from the former natural forest cover: this is most likely with the woods in steep-sided valleys or gorges and mixed deciduous woodland such as Roudsea (Birks 1982), where pollen analysis suggests considerable continuity in woodland composition over the last 5000 years. Some of the pure sessile oak woods are old secondary woodland (but may still be ancient if their origin is pre-1600); in other, formerly mixed coppices, the oak has been selected for or enhanced by planting.

9.4 How many woodland sites, species and communities should be conserved?

Conservation must start from the identification of the sites, and of the species and features within them, that are important. There must be systems for ensuring that the subsequent management of these sites can be influenced so that their nature conservation value is maintained. Finally, there must be clear guidance as to what types of management are appropriate on any given site.

Cumbria contains 54 450 ha of woodland (Forestry Commission 1984). Of this, 15 880 ha (about 29%) is thought to be ancient woodland, of which 10 518 ha is thought to be ancient semi-natural woodland (Figure 1) (Whitbread 1985). These ancient woods are likely to contain the majority of sites important for nature conservation. More detailed field surveys have been used to select Sites of Special Scientific Interest (SSSIs) and National Nature Reserves (NNRs), but this process is by no means complete.

In the small proportion of woods that become NNRs (Figure 2), the NCC can demonstrate particular types of management for nature conservation, including minimum intervention treatment. This last may be difficult for other owners to justify or afford. The NCC has a statutory right under the Wildlife and Countryside Act 1981 to comment on an owner's management plans for a wood designated as an SSSI. Compensation can be paid for loss of profit that results from any management recommendations that the NCC makes. Elsewhere, the NCC has no specific

References

Birks, H.J.B. 1982. Mid-Flandrian forest history of Roudsea Wood National Nature Reserve, Cumbria. *New Phytologist*, **90**, 339–354.

Brooks, A. 1980. *Woodlands*. Wallingford: British Trust for Conservation Volunteers.

Forestry Commission. 1984. *Census of woodlands and trees, county of Cumbria*. Edinburgh: Forestry Commission.

Forestry Commission. 1985a. *The policy for broadleaved woodlands*. (Policy and procedure paper no. 5.) Edinburgh: Forestry Commission.

Forestry Commission. 1985b. *Broadleaved woodland grant scheme*. Edinburgh: Forestry Commission.

Forestry Commission. 1985c. *Guidelines for the management of broadleaved woodland*. Edinburgh: Forestry Commission.

Forestry Commission. 1985d. *Wildlife Ranger's handbook*. Edinburgh: Forestry Commission.

Grant, W. 1981. Management and development strategy in a multiple use forest. In: *Research and planning for nature conservation and amenity in woodlands*, edited by K. Hearn, 50–70. Proceedings of the Recreation Ecology Research Group Meeting.

Halliday, G. 1978. *Flowering plants and ferns of Cumbria*. (Centre for North-West Studies occasional paper no. 4.) Lancaster: University of Lancaster.

Hill, M.O. 1979. The development of a flora in even-aged plantations. In: *The ecology of even-aged plantations*, edited by E.D. Ford, D.C. Malcolm & J. Atterson, 175–192. Cambridge: Institute of Terrestrial Ecology.

Hill, M.O. & Jones, E.W. 1978. Vegetation changes resulting from the afforestation of rough grazings in Caeo Forest, south Wales. *Journal of Ecology*, **66**, 433–456.

Kirby, K.J. 1984. *Forestry operations and broadleaf woodland conservation*. (Focus on nature conservation no. 8.) Peterborough: Nature Conservancy Council.

Miles, J. & Kinnaird, J.W. 1979. Grazing with particular reference to birch, juniper and Scots pine in the Scottish Highlands. *Scottish Forestry*, **33**, 280–289.

Moss, D. 1979. Even-aged plantations as a habitat for birds. In: *The ecology of even-aged plantations*, edited by E.D. Ford, D.C. Malcolm & J. Atterson, 413–427. Cambridge: Institute of Terrestrial Ecology.

Nature Conservancy Council. 1984. *Nature conservation in Great Britain*. Peterborough: Nature Conservancy Council.

Nature Conservancy Council. 1986. *Nature conservation and afforestation in Britain*. Peterborough: Nature Conservancy Council.

Perring, F.H. & Walters, S.M. 1982. *Atlas of the British flora*. 3rd ed. Wakefield: EP Publishing Ltd.

Peterken, G.F. 1977. Habitat conservation priorities in British and European woodlands. *Biological Conservation*, **11**, 223–236.

Peterken, G.F. 1981. *Woodland conservation and management*. London: Chapman & Hall.

Peterken, G.F. 1983. Woodland conservation in Britain. In: *Conservation in perspective*, edited by A. Warren & F.B. Goldsmith, 83–100. Chichester: Wiley.

Ratcliffe, D.A. 1968. An ecological account of Atlantic bryophytes in the British Isles. *New Phytologist*, **67**, 365–439.

Ratcliffe, D.A., ed. 1977. *A nature conservation review*. Cambridge: Cambridge University Press.

Smart, N. & Andrews, J. 1985. *Birds and broadleaves*. Sandy: Royal Society for the Protection of Birds.

Sykes, J.M. 1981. Monitoring in woodlands. In: *Forest and woodland ecology*, edited by F.T. Last & A.S. Gardiner, 32–40. Cambridge: Institute of Terrestrial Ecology.

Whitbread, A.M. 1985. *Cumbria inventory of ancient woodland* (provisional). Peterborough: Nature Conservancy Council.

Woodland amenity and recreation in the Lake District

A Fishwick

Lake District National Park Authority, Kendal

10.1 Introduction

Amenity values cannot be quantified in terms of annual increment or compared using net discounted revenues, but few people would deny that trees and woodlands make an enormous contribution to the beauty of the Lake District. Recreational use can be measured, but the information currently available gives little more than a broad impression of the extent to which woodland is used for different purposes. Surveys indicate that most visits to woodlands in the Lake District are an incidental part of a trip or walk in the National Park, which tend to focus on the villages, lakeshores and fells (Fishwick 1985).

I propose first to explore the amenity and recreational values of the woodlands of the Lake District, and then to discuss the statutory powers and mechanisms that have been developed in response to the needs to protect and secure those values.

10.2 Trees and woodlands in the landscape

Trees may form the centre piece or frame for much-loved views (the Scots pine (*Pinus sylvestris*) on Friar's Crag, Derwent Water), a foil for buildings (the characteristic grouping of trees sheltering Lake District farmsteads) and a screen for development judged best hidden (caravan sites and car parks). Scattered small woods, of which there are many, and more extensive areas, whether coniferous (Grizedale Forest) or broad-leaved (Rusland Valley), obviously have a considerable influence on the appearance of the areas in which they occur. To assess their contribution requires a willingness to trade in value judgements which are invariably deeply rooted.

Landscape tastes

In the Lake District National Park, the intensively humanized patchwork of enclosed farmland is set against the backdrop of the wider fells. Nicholson (1959) has traced the way in which the perception of mountain scenery changed in Britain in the 18th century: from being regarded as nature's 'shames and ills', mountains were transformed to 'temples of nature built by the almighty'. That transformation was consolidated by the romantic descriptions in Wordsworth's writings and poetry.

Today the exceptional combination of natural and cultural elements is held in sufficient regard nationally for the government to have submitted the National Park for inclusion in the list of World Heritage Sites in 1985.

The Manchester Corporation Waterworks Act of 1879, relating to the creation of the Thirlmere reservoir, contains provisions to safeguard access 'heretofore actually enjoyed on the part of the public and tourists to the mountains and fells' and to require the planting of indigenous species on the reservoir margins. The Act illustrates quite clearly that concerns about recreation and amenity are long standing.

It is worth remembering that some at least of the Lake District woods were planted for amenity purposes using the wealth of the industrial revolution.

The agreement drawn up in 1936 between the Forestry Commission (FC) and the Council for the Preservation of Rural England remains a landmark in the controversy surrounding afforestation in the National Park. Both parties agreed that 'there is in the Lake District a central block, which by reason of its unusual beauty and seclusion and its remoteness, should be ruled out altogether from afforestation by the Commissioners'. Protests about afforestation provided one of the *foci* for the campaign for the establishment of National Parks in England and Wales, and the literature associated with that campaign and official government reports of the period (1930–49) show prevailing attitudes to woodlands and afforestation which still have considerable influence today.

Forest gloom: woodland glory

One of the first plans covering the Lake District (Abercrombie & Kelly 1932) drew attention to the undesirable loss of native deciduous woodland. 'The normal English countryside may be said to owe a great deal of its mature and cultivated beauty to the careful planting which was largely done in the seventeenth, eighteenth and early years of the nineteenth century . . . on the other hand planting can cause serious disfigurement when a bare and austere scene is reduced in stature by serried rows of conifers . . . much of the objection to this commercial planting would disappear if it were possible to plant deciduous trees instead of conifers.'

Symonds (1936) described Whinlatter and Ennerdale Forests thus: 'What is seen is the rigid and monotonous ranks of spruce, dark green to blackish, goose-stepping on the fellsides. Their colour – you must except the larch – is in effect one steady tone all round the dull year: there are no glories of spring and autumn for the conifer. Sunlight, which a broadleaved deciduous tree reflects and vivifies, is annihilated on their absorbent texture: on a bright landscape they are so much blotting paper . . . with the sitka spruce king of this gloomy kingdom, you must add the curse of uniformity in growth . . . they have the air of mass production, the efficiency of the machine.' In a recent forestry seminar, A Mattingley of the Ramblers' Association argued that this denunciation of the effects of commercial afforestation on the landscape has never been bettered.

In the seminal report on the concept of National Parks in England and Wales (Ministry of Town and Country Planning 1945), Dower echoes Symonds' distinction between the 'gloom' of extensive coniferous plantations and the 'glory' of the deciduous woodlands. Dower listed 'large-scale afforestation, blanketing the varied colours and subtle moulding of the hillsides with monotonous sharp-edged conifer plantations' and 'ill considered felling of woodlands or hedgerow timber of amenity value' amongst a catalogue of 'misuses and disfigurements' which threatened the integrity of National Park areas, and which he felt ought to be controlled. In contrast, he cited 'the intricacy with which the fields and coppice woods are often intermingled in the Lake District valleys' as an example 'of our most beautiful countryside' in need of positive measures for their conservation.

The condemnation of conifers was strong in official reports, but not absolute. Dower (Ministry of Town and Country Planning 1945), for example, regarded the landscape of Grizedale 'better suited to large scale afforestation' (than the central Lake District), and in references to amenity planting he wrote that 'a judicious mixture of larch (Larix spp.) and scots may do no harm' but 'all other conifers are best avoided'.

Coniferous afforestation has also been criticized for its adverse effect on public enjoyment because it interferes with views and places limitations on access. The first review of the Lake District's development plan (Lake District Planning Board 1965) commented: 'The incentive to abandon the freedom of an open fell for the sepulchral glooms of a coniferous plantation must normally be small . . . wherever fell slopes, however featureless, are taken for such planting something is lost in terms of landscape enjoyment'. However, it is a

fact that the FC has done a great deal to provide for visitors in its forests. The innovative work of Chard (1967) and Grant (1971) at Grizedale has provided a model for developments elsewhere in the country. There have also been considerable improvements in forest design.

The Hobhouse Report (Ministry of Town and Country Planning 1947), which made recommendations on the selection of National Parks and measures to secure their protection, recognized the way in which nature conservation and landscape values converged in the broadleaved woodlands: 'Trees and woodlands, and the wealth of woodland flora and fauna, will contribute so much to the beauty and interest of the Parks that their preservation and maintenance must be a vital factor in National Park planning'. However, it is only recently that this convergence of values has been more fully recognized in both local (Fishwick 1981) and national (Forestry Commission 1984) policies, which have stressed the importance of retaining existing semi-natural broadleaved woodlands.

Views of the public

The results of two consultation exercises undertaken in the Lake District National Park during the last ten years provide some insights into current attitudes towards forestry. In 1976 the Lake District National Park Authority (NPA) published *Ideas for discussion*, a series of leaflets designed to provide statutory and voluntary bodies and individual members of the public with an opportunity to comment on proposed policies for the management of the National Park. More recently, in 1985, a lower-key exercise was conducted using a short questionnaire based on a leaflet *Focus on the Lake District National Park*.

Over 500 individual responses were recorded against each of the suggestions in the *Forestry and woodlands* leaflet of the 1976 series, and additional comments were received from 36 organizations (Lake District Special Planning Board 1976). Suggestions to retain valued elements of the landscape were given strong support (only 1–4% of respondents expressing disagreement with related policies), but *caveats* reflected concern that the economics of the situation must be considered. A suggestion that there should be a strong presumption against large-scale afforestation met with a higher level of disagreement (11% of respondents). Adverse comments were, however, split between those who said there was no scope for large-scale afforestation in the Park, and those who accepted more forestry to support employment and to contribute to the nation's timber resources.

The *Focus* leaflet issued in 1985 outlined the main elements of the first National Park plan – basically, that the Park is not an appropriate place for further large-scale expansion of afforestation, but new planting of a scale and type which would not spoil the individual character of the area is acceptable; and that the NPA supported the conservation of the broadleaved woods, one of the Lake District's most important scenic and wildlife assets. The brief commentary also covered policies on farming, the debate about the future of the Common Agricultural Policy and the need to encourage farming as an important part of the Lake District economy on which much of the landscape depends. At the end of two short paragraphs, the question was posed: 'What are your views on the future of farming and forestry in the National Park?' The number of responses received was 334 (Lake District Special Planning Board 1985a), and a summary of them is given in Table 1.

Table 1. Summary of the farming and forestry responses of the 1985 survey

Response	Number of responses
Policies should be kept as they are	23
In support of the 1978 plan	16
Against any major expansion of forestry	21
No further establishment of conifer plantations	27
Conifer planting should be limited or severely limited	25
Greater control over forestry	10
Planning control over forestry	8
More forestry to support the local economy and local employment	14
A limited expansion of forestry	14
More forestry, using mixed species and with attention to design and location	27
More broadleaved planting	45
A preference for broadleaves over conifers	27
Conservation of existing broadleaved woodland	21
Farming should be supported to maintain the landscapes or local employment and local communities	160

The approach taken in these two exercises, the way in which questions and supporting text can tend to foster particular answers, and the representatives or otherwise of those who actually respond are all open to criticism. Nevertheless, the general tenor of the views of those people willing to put pen to paper suggests

that the basic distinction in attitudes towards conifers – associated with extensive afforestation – and broadleaves still prevails. The response favouring more broadleaved planting may have been enhanced by the recent publicity about the conservation of broadleaves. However, individual cases, such as the campaign against the FC's proposals for afforestation at Grassguards in Dunnerdale which reached the national press, show that protests against afforestation with conifers in the Lake District still have strong public appeal.

While additional afforestation draws protests, it is interesting that the amenity bodies that are largely responsible are also expressing their support for the FC's continuing management of existing forests. Selling off the public forest estate to the private sector is seen as a threat to the continuation of landscape improvements in established plantations, and to the progressive conservation and recreation policies developed by the FC.

10.3 Recreation in forests and woodlands

Severn (1942) described the Lake District as 'a rough, shooting country, and strictly speaking a bad game country . . . a distinctly good woodcock country', but over many years the use of woodlands in the Lake District has changed from an emphasis on shooting to encompass a variety of recreational and educational activities. This change partly reflects a diversification in the pattern of woodland ownership, with estates and farms being joined by conservation and amenity bodies and educational establishments. Also, there have been new demands, eg for orienteering, and new initiatives in response to the growing public interest in the historical and cultural legacy of the coppice industries.

Forestry Commission's recreation provision

The FC manages over 9600 ha of forests and woodlands in the Lake District and within them offers a wide range of recreational opportunities. It has two visitor centres, Grizedale and Whinlatter, which provide *foci* for forest walks and permanent orienteering courses. Permits are arranged for activities as diverse as stalking, fishing, pony trekking, car rallies, orienteering events and the use of hides for wildlife observation. Buildings have been let to provide outdoor activity centres, and provision has been made for local schools to use forest plots for educational purposes. At Grizedale,

Plate 1. Long How Wood, Buttermere. Sheep grazing in deciduous woodlands leads to impoverished ground flora and poor tree regeneration. (Photograph J K Adamson)

Plate 2. A weir on a drainage ditch at Kershope Forest. Joint research work by the Forestry Commission and the Institute of Terrestrial Ecology is leading to an understanding of the effects of coniferous forests on drainage water. (Photograph J K Adamson)

Plate 3. The recently restored Stott Park Bobbin Mill at Finsthwaite which is open to the public. Increasing interest in ancient wood-processing industries may have important local influences on woodland management. (Photograph J K Adamson)

Plate 4. Radnor in Wales. Excess agricultural production and high timber imports may result in the introduction of agro-forestry to Cumbria. A joint (FC, IGAP, MLURI) agroforestry research site where larch trees (*Larix* spp.) are widely spaced to allow undergrazing by sheep. (Photograph G J Lawson)

a field is let to the Camping Club for a 60 pitch camp site, and three back-packers sites have been made available elsewhere.

The FC has provided some 27 car parks, ranging in size from quite small lay-bys to large car parks of 75 or more spaces. Individual forest maps have been published to make the forests more accessible to the public. At Grizedale, in conjunction with a local Trust, the FC has supported the establishment of a theatre and is currently contributing to a sculpture project. As a result, a sculpture trail has been added to the more conventional forest walks.

Information on levels of use is limited. It is estimated that 100 000 people make use of the facilities at Grizedale each year (J C Voysey pers. comm.) and the visitor centre at Whinlatter attracts some 70 000 each year (M Scott pers. comm.). These figures need to be seen in perspective. Very popular recreation sites in the Lake District, such as Tarn Hows and White Moss Common, are thought to have in excess of 500 000 visitors each year. Although woodlands are an important component of both these areas, their popularity owes much to other factors – accessibility, water, viewpoints and a long history of recreational use.

Outside the FC's holdings, there is little formal provision for recreation in woodlands, although many woodlands are served by public rights-of-way or permissive paths. The few nature trails appear not to have attracted large numbers of people, although annual shows of daffodils (*Narcissus pseudonarcissus*) or bluebells (*Hyacinthoides non-scripta*) may draw visitors back regularly to particular sites. For the most part, woodlands are enjoyed by millions each year as a part of the Lake District, rather than as an end in themselves, although there are significant exceptions to this generalization.

Orienteering

Orienteering is a sport which has grown significantly since the 1960s. The Lake District provides some excellent orienteering country and seven clubs, covering an area from the Scottish Borders to Merseyside, have mainly been responsible for mapping suitable areas. Over 40 areas, some overlapping, have been mapped for orienteering covering some 150 km^2 (M Collett pers. comm.), with more in the pipeline for forthcoming events. A good deal of orienteering in the area takes place on open land so just over 40% of the map-

ped area is woodland or forest. Levels of use for particular events vary. A club training evening may attract 50 or more people and a local competitive event 200; badge events which enable competitors to achieve times contributing to nationally recognized awards typically may have 500–700 entrants; and national ranking events may well attract over 1000.

The number and status of events in any one year vary. Major national events occur irregularly, and at present there are no more than a handful of the badge and ranking events each year. Even taking account of training, few sites are used more than two or three times annually by the clubs.

Youth group activities

The development of outdoor activity centres and the use of the Lake District by school and youth groups for adventure education and field studies has been a major growth area. In the early post-war years, a handful of centres were established, amongst them the Outward Bound Schools, Brathay and the YMCA's national centre at Lakeside on Windermere. The Scout and Youth Hostels movements also became well established. Today there are 27 staffed centres, nearly 90 unmanned or supervised huts, 23 Youth Hostels, seven similar hostels run by other organizations, and at least 13 guest houses and hotels taking youth and school groups fairly regularly. The Scouts have 57 sites which they use for camping, and another 59 camp sites are known to take youth groups.

The Brathay woodlands project tried to establish the nature of the use and interest in woodlands by such centres and groups. Preliminary questionnaire surveys revealed a wide range of uses – for rope-ways, initiative exercises, camping, shelter building, orienteering, nature studies, and walks (M Gee pers. comm.). The study also drew attention to the fact that several of the centres own significant areas of woodland, providing scope for integrating course activities with the management of the woodlands.

The woodland scene and interpretation

The interest that has been fostered in broadleaved woodlands nationally, and in particular in the significance of the semi-natural woodlands and their history, is reflected in a number of new initiatives in the Lake District. These have added to the existing interpretative work of the FC and educational centres such as Haybridge. In 1981, the NPA took a lease on the

Duddon Iron Furnace and has been consolidating the substantial remains. In 1983, English Heritage opened Stott Park Bobbin Mill with working machinery, and in 1984 the embryo New Woodmanship Trust held a 'Weekend in the woods' at Brantwood on the shores of Coniston, attracting a thousand visitors to a mix of events, including a charcoal burn, craft demonstrations advertising examples of products which can still be produced from the woodlands, and woodland walks. The Trust is particularly interested in novel approaches to the management of woodlands involving the community, voluntary societies, co-operatives, and visitors to the countryside – a merging of action and interpretation.

10.4   Woodland amenity and development

Development

Perhaps the most 'financially productive' use of woodlands (at least of those that are unlikely to blow down too easily!) is to fill them with caravans or time-share units. If the ability to hide development was the most significant criterion in determining applications for development, undoubtedly more woodlands would have been exploited. Many other factors are taken into account. Nevertheless, a significant amount of woodland is occupied by such sites.

From a fairly rapid survey of small-scale tourist maps, it is thought there are at least 15 large caravan or chalet sites (more than 20 units each) developed in woodlands. These sites, together with camp and outdoor activity centres, such as the Great Tower Scout Camp, occupy in excess of 200 ha of woodland. In addition, there are at least 20 woods containing reasonable-sized car parks (more than 25 cars). Many other caravan sites and car parks are effectively screened by small trees and small woods.

In these situations there is considerable concern that the screening should remain effective in the long term, but also there is often concern about public liabilities which result in the felling of aged trees and the removal of branches showing signs of dying back. Most of these facilities have been established over the last 25 years, and in the years ahead a considerable amount of specialized woodland management will be required.

Both the retention of existing trees and the planting of new ones play a significant part in integrating existing buildings and new development in the landscape. The vast majority of the 123 Tree Preservation Orders (TPOs) confirmed in the Lake District are associated with this aspect of the NPA's work. TPOs cover over 550 trees, plus nearly 100 small groups of trees or small woods, which are associated with existing buildings, in the open countryside, on the edges of hamlets and villages, or on new development sites. Landscape conditions requiring the retention of existing trees or new planting are also used to effect, when approving applications for development. Since 1980, 14 Conservation Areas have been designated and these also give added statutory protection to existing trees in such areas.

Carrying capacity

The fact that woodlands are capable of 'hiding' developments has led to a conventional view that they have a high carrying capacity, that is they can absorb a great deal of recreational use without detriment, but this assessment appears to be based on the view from a distance. It is evident that, wherever there is public access on any scale in woodlands, there are localized problems of erosion of the ground and a general tendency for apparently random networks of paths to develop. Questions have also been raised about disturbance to wildlife.

There has been considerable concern about the impact of large orienteering events. What little monitoring has been done suggests that single events do relatively little physical damage, except perhaps across boggy areas. Physical damage is only one criterion, and it is a fact that existing orienteering maps include a few areas with uncommon or rare species which are vulnerable to trampling. Disturbance to breeding sites of birds and deer is another issue about which little is known (whether caused by orienteering or informal recreation). However, with an organized sport such as orienteering, it is possible to minimize problems by avoiding sensitive times of the year on particular sites and demarcating 'no go' areas. The regular use of individual sites is a different proposition, and there is evidence, from permanent orienteering courses and outdoor centres, that initial trampling may lead fairly quickly to more permanent tracks and erosion scars.

10.5   Powers and mechanisms

Legislation 1949–85

The National Parks and Access to the Countryside Act 1949 placed on those local authorities responsible for

National Parks a general duty to formulate and carry out proposals for the purpose of preserving natural beauty and promoting the Parks' enjoyment by the public. Local authorities were also given a general power to plant trees for the purpose of preserving natural beauty or improving the appearance of derelict land in their area. Specific powers in the 1949 Act were concerned with the provision of accommodation, camping sites and parking places, and with securing access to waterways and open country. Open country was defined as mountain, moor, heath, down, cliff or foreshore, and the legislation enabled authorities to make access agreements and orders or to acquire land compulsorily. In deference to the priority given to post-war timber production, there were provisions to exempt from access orders land used, or about to be brought into use, for growing timber for commercial purposes.

The Countryside Act 1968 modified these provisions. Picnic sites were added to the list of facilities, and the definition of open country was extended to include woodlands. The exclusion of woodlands from access orders was modified to the effect that the prejudicial effect on timber production had to be shown to outweigh the benefits arising from facilities for access. Section 11 of the 1968 Act charged every minister, government department and public body to have regard to the desirability of conserving the natural beauty and amenity of the countryside in the exercise of their functions. Section 23 enabled the FC to provide tourist, recreational and sporting facilities on land that it managed. The FC's responsibilities for conserving natural beauty were made more explicit in the Wildlife and Countryside (Amendment) Act 1985.

Section 39 of the Wildlife and Countryside Act 1981 also clarified National Park authorities' powers to make management agreements for conservation purposes as well as access, and Section 43 required them to prepare maps of moor and heath. Under Section 3 of the 1985 (Amendment) Act, the maps were extended to include woodlands and coastal features.

The Hobhouse prescriptions

This brief summary of statutory powers can be compared with proposals made by the Hobhouse Committee (Ministry of Town and Country Planning 1947). The Hobhouse Committee saw the tree preservation procedure, then a part of a new planning bill, as a means of protecting privately owned amenity woods and trees.

It also recommended that Park committees should be able to specify areas in the National Parks within which proposals for the planting of new woodlands should be subject to their consent.

The Hobhouse Committee considered that its proposed National Parks Commission should have powers to enter into agreements with landowners covering the felling, planting or management of woods and to purchase (by agreement, or in rare cases, compulsorily) existing woodlands of special beauty or sites for new planting. The Committee hoped that conflict between national programmes of timber production and National Park objectives would be avoided by referring all plans for dedicated woodlands and for land acquisitions by the FC to the National Parks Commission, so that any reasonable requirements to promote or safeguard amenities could be considered properly.

The framework of arrangements proposed by Hobhouse has been followed in broad terms. Responsibilities for consultations and agreements have rested with local rather than national bodies, but, as already noted, the earlier legislation was concerned much more with access than with conserving natural beauty. Efforts to secure formal control of new planting have been resisted by landowners, the FC and government, but maps have played a part in the development of forestry policy in National Parks. However, aspects of the arrangements have been criticized and are under review at present.

Current practices and issues

The role which TPOs and other sections of the planning legislation play in protecting trees on development sites and in Conservation Areas has already been mentioned. At the present time, the government is considering the introduction of a Landscape Conservation Order. Various models have been considered (Council for National Parks 1985), but the one most favoured by the National Parks and amenity bodies would take the form of a stop order to prevent destruction of an important landscape feature or habitat, followed by a period of negotiation to agree management proposals to ensure the feature is preserved. Payments would be involved, but it is hoped that such an order would form a more constructive option to the negative instrument of the TPO when, say, woodlands are concerned, and would resolve uncertainties over compensation.

Consultation arrangements enable the NPA to try to resist undesirable planting, to influence the design and species mix of new planting, and to modify proposals for felling and woodland management. The arrangements cover revision of plans of operations for the remaining dedicated woodlands, forestry grant schemes, felling licences, and the FC's proposals for its own woods and plantations. The system of *ad hoc* arrangements was codified in 1974, when a disputes procedure involving the FC's Regional Advisory Committees (RACs) was introduced. Consultations on Ministry of Agriculture, Fisheries and Food (MAFF) grant-aided shelterbelt schemes are covered by consultation arrangements with farmers on all grant-aided work in the National Park, introduced in 1981.

For the year April 1985 to April 1986, the NPA dealt with 35 consultations with the FC, mainly for forestry grant schemes and felling licences, and 17 MAFF grant notifications for proposed shelterbelts.

Since the forestry consultations were formalized in 1974, the NPA has only referred two schemes for afforestation to the RAC, having failed to reach a satisfactory conclusion with the applicants. On each occasion – Hudson's Allotment and Thornholme Farm – the NPA found it necessary to protest about the manner in which such cases are conducted (Lake District Special Planning Board 1985b). The NPA is most likely to refer a case to the RAC when a matter of principle is at stake: in both these cases, afforestation of open fell. The RAC is required to try to find an accommodation between the parties in dispute, rather than adjudicate on the merits of respective cases. Where no accommodation can be found, the case is referred to the Forestry Minister, who is advised by the FC Conservator of cases submitted and of the views of the RAC.

This system has drawn considerable criticism, which was acknowledged by William Waldegrave in the government's response (Hansard 1985) to the uplands debate (Countryside Commission 1984) and, as a result, the FC has reviewed the arrangements. The conclusions of the Forestry Commissioners, circulated recently in a consultation paper (Forestry Commission 1986), appear to have failed to tackle the fundamental criticisms lodged by amenity bodies, who continue to campaign for planning control.

The existence of the consultation arrangements was used by government (Department of the Environment 1976) to reject the recommendations of the Sandford

Committee (Department of the Environment 1974) that afforestation in National Parks should be subject to planning control, and more recently to reject a similar argument put by the Countryside Commission in conclusion to the uplands debate (Hansard 1985). However, it seems likely that, if cases of afforestation of land without FC grant, and thereby without consultations with local authorities, continue to occur, the government is likely to have difficulty in resisting public demands for some form of control.

Forestry maps

In 1961, a voluntary agreement was announced between the Timber Growers' Organization, the Country Landowners' Association, the FC and the then National Parks Commission for the preparation of forestry maps and improved consultations. The agreement coincided with considerable concern about afforestation on Exmoor and Dartmoor (Standing Committee on National Parks 1961). The National Parks Commission undertook to advise Park committees to produce maps showing areas of land in the Parks where:

i. afforestation was totally undesirable;

ii. afforestation might be accepted under varying circumstances;

iii. afforestation could be accepted.

The Lake District NPA declined to produce a map modelled on this agreement, but, through the efforts of C D Acland, then the National Trust's Regional Agent and a member of the Lake District Planning Board, agreed an alternative set of consultation arrangements. These were eventually succeeded in 1974 by the national scheme. The Board's policies did, however, hinge on the existence of the 1936 Agreement, but it was stressed that any increase in afforestation other than of a strictly limited nature outside the central area and dales would be strongly opposed (Lake District Planning Board 1965). Policy maps for both afforestation and broadleaved woodlands were introduced in the first National Park plan (Lake District Special Planning Board 1978a).

In 1981, a mapping requirement was placed on the National Park by Section 43 of the Wildlife and Countryside Act. This arose from the continuing controversy surrounding the ploughing of moorland in Exmoor. As already noted, the map has been extended by the Wildlife and Countryside Amendment Act 1985 to include woodlands. It has to be prepared in accordance with guidelines to be produced by the Countryside

Forestry and water quality

M Hornung[1] and J K Adamson[2]

Institute of Terrestrial Ecology, [1]Bangor, [2]Grange-over-Sands

11.1 Introduction

Interpretation of the term 'water quality' varies with geographical location and the use to be made of the water. Different criteria will be applied by fishermen, water supply engineers and industrial plant operators, for instance. In this paper, we attempt to take a wide view of the term and we consider water quality as the chemistry, colour and sediment content of streams, rivers and lakes.

The impact of forestry on water quality can be seen as the result of two sets of interacting processes: those which are due to the growth of trees, as opposed to other vegetation types, and those due to forest management practices. Trees, as a result of their tall growth form, 'capture' more particulate and aerosol material from the atmosphere than low-growing vegetation types. Moisture losses through evapotranspiration are greater from forests than grassland, agricultural crops or moorland, especially at higher altitudes. The root systems of trees can create a network of macro-pores which act as pathways for rapid water movement to depth. The surface soil horizons developed below forests differ, both chemically and physically, from those formed under other vegetation types – this difference can influence water pathways and chemistry. Forestry management practices which can influence water quality include ploughing, drainage, application of fertilizers, use of pesticides, creation of forest roads, harvesting and timber extraction. Whether or not these practices are used, and, if used, the intensity of their use will vary with location and conditions. On the more fertile, better-drained soils of the lowlands, cultivation, drainage and fertilizer applications may not be required. Selective harvesting may be used in broadleaved woodlands, while clearfelling is normally used in coniferous forests. Herbicides may be needed to suppress competing vegetation on some sites but not on others.

In the lowland areas dominated by intensive agriculture, drainage waters from woodlands generally have lower contents of sediments and lower concentrations of nitrate, phosphate and calcium than streams draining agricultural land. This situation reflects the greater inputs of fertilizers and the more frequent ploughing of agricultural land. In this intensive agricultural zone, coniferous and broadleaved woodland is often seen as protecting water quality, limiting erosion, and preventing pollution by fertilizers or pesticides. In upland areas dominated by extensive sheep farming, forestry may

represent the more intensive land use, particularly as ploughing, drainage and fertilizer additions may be needed to give satisfactory forest growth on the acid, poorly drained soils. Upland rivers and streams commonly have low contents of solutes, so that even a small change in solution chemistry, as a result of forest activities, may be significant.

In this paper, we consider the likely impact of the main forest management activities/stages on water quality and the link between forestry and the acidification of streams.

11.2 Site preparation

Afforestation in the uplands, particularly on soils grouped as stagnopodzols, stagnohumic gleys and peats (Avery 1980), generally involves ploughing and drainage. The ploughing (Thompson 1984) is designed to improve rooting conditions: it increases aeration and temperature, improves drainage, leads to increased rates of organic matter breakdown, and provides an elevated planting position. Ross and Malcolm (1982) studied the impact of ploughing on a peaty ferric stagnopodzol soil in south-east Scotland: the tilled soil had a lower bulk density, was better aerated, showed faster infiltration and had higher mean annual temperatures than untilled soil. The tilled soil also showed intimate mixing of the organic and mineral horizons to a depth of 60 cm. Drainage ditches, linked to the plough furrows, are intended to remove surface water rapidly (Thompson 1979). Drainage produces some disruption of soil horizons, exposes subsoil and drift material, and deposits subsoil material on the surface alongside the ditch. The combined results of the ploughing and drainage can produce changes in drainage water quality.

Robinson (1980) examined the changes in hydrology and water quality during site preparation in a catchment drained by the Coal Burn, a headwater tributary of the River Irthing. Although located just over the county border, in Northumberland, the site is typical of much of northern Cumbria. The vegetation was dominated by purple moor-grass (*Molinia caerulea*) on peats and stagnohumic gley soils and underlain by glacial till. Concentrations of suspended sediments in the stream increased as a result of the ploughing and drainage, peaking during and immediately after the operations, and then declining. Before site preparation began, sediment concentrations ranged from 1 mg l^{-1} to 28 mg l^{-1} with a mean of 3.6 mg l^{-1}. During drainage operations, concentrations reached a peak of 7720 mg l^{-1} and a mean of 207 mg l^{-1}. Most of the very high concentrations occurred towards the end of the operations, when link and cross drains were being dug and when a bucket drainer was used to widen part of the main channel. Seven years after drainage, the sediment concentrations ranged from 1 mg l^{-1} to 5 mg l^{-1} in low flows and up to 40 mg l^{-1} in floods; the discharge weighted mean was 15 mg l^{-1}. Burt, Donohoe and Vann (1983) reported a similar increase in sediment yields during forestry drainage operations at a site in the southern Pennines where the soils were peaty gleys, podzols and brown soils. The maximum sediment concentrations recorded were 5384 mg l^{-1}, and total sediment yields were large enough during a storm to pollute a local reservoir. At most sites, increased sediment levels seem to decline rapidly after ploughing and drainage, but Newson (1980) has shown that they can continue for at least 40 years in certain drift materials, due to erosion of drainage channels.

Robinson (1980) also reported increased solute concentrations of calcium (Ca), magnesium (Mg), nitrogen (N) and potassium (K) following drainage. There was also a change in the relative abundance of the four main cations, from Na>Ca>Mg>K before drainage, to Ca>Na>Mg>K after drainage. Two years after cultivation and drainage, solute concentrations were similar to pre-treatment levels. Robinson suggests that the changes in solute concentrations were probably due to the exposure of till in the drains. Drainage of peats can also produce increased concentrations of ammonium and nitrate, due to increased mineralization of organic nitrogen compounds.

The increased solute concentrations will have a very limited impact; they are not high enough to necessitate additional treatment of water supplies for public use, and downstream dilution would rapidly take place. They are also unlikely to affect freshwater biota directly. The increased sediment yield may be a greater problem. It can lead to pollution and sedimentation of adjacent small reservoirs and damage to fish spawning beds. Road building can also produce large increases in sediment yield, and caused a major pollution problem at the Cray Reservoir in south Wales (Stretton 1984).

The impact of site preparation on water quality can be reduced by the careful planning of drainage schemes and by minimizing cultivation. The drainage schemes used by the Forestry Commission today are very different from those used 30–40 years ago. Thus, the design of drainage schemes used at the sites referred to in these examples are unlikely to be used today: these sites can be seen as 'worst possible cases'. New approaches to cultivation and drainage are giving very encouraging results; at a study site near Llanbrynmair, in mid-Wales, the increase in sediment yields during ploughing and drainage were insignificant (M D Newson pers. comm.). Research is also continuing on the potential of techniques such as mole-drainage, and in Cumbria scarifiers and rippers are beginning to replace ploughs. The careful planning and siting of roads are also reducing their impact.

In lowland areas or on better-drained, more fertile soils, the amount of site preparation may be minimal. Ploughing and drainage may not be necessary and the planting will, therefore, have little or no impact on water quality. Similarly, restocking existing woodland or plantations generally involves little ground disturbance, although the cleaning of existing drains may have a short-lived impact on water quality.

11.3 Fertilizers

Fertilizers have been widely used in forestry on the less fertile soils and particularly, therefore, in the new forests planted on the acid soils of the uplands. Phosphorus has been, and still is, the most widely used fertilizer (Mayhead 1976); it is generally applied as rock phosphate and at rates of about 50 kg P ha^{-1}. As described by Voysey (see page 18), application may be at establishment or at pole stage, or both. Nitrogen, as prilled urea or ammonium nitrate, has been used on heather (*Calluna vulgaris*)-dominated sites planted with Sitka spruce (*Picea sitchensis*): rates of application are usually around 150 kg N ha^{-1}. However, tree species mixtures, as described by Brown and Dighton (see page 65), are becoming increasingly attractive for overcoming nitrogen deficiencies, particularly because of the high cost of nitrogen fertilizers. Controlling the competing heather is another option. Potassium, as potassium chloride, is used at the establishment phase on deep peats at rates of approximately 100 kg K ha^{-1}. Fertilizer applications at establishment are often (but not solely) by hand, while aerial application is now most common for additions at pole stage. If any of these fertilizers find their way into drainage waters, they will clearly have an impact on water quality.

Nutter (1979) has suggested that, on a world-wide basis, application of fertilizers to forests has little impact on water quality when proper safeguards are used, and in the absence of overland flows of water. The direct input of fertilizer to streams during aerial application is said to be often the sole reason for fertilization affecting streamwater quality. Both Nutter (1979) and Tamm et al. (1974) do, however, stress that care is needed when using readily leachable forms of nitrogen. Despite comments of this nature, there is considerable concern in the UK water industry about the possible impact of forest fertilization, particularly phosphorus, on the chemistry of otherwise unpolluted streams and rivers in the uplands (Youngman & Lack 1981). A small increase in phosphorus levels in these waters can lead to a large increase in biological production, as the biological populations in the streams, lakes and reservoirs are generally phosphorus limited. The water industry's main concern is with reservoirs, where a build-up of phosphorus, leading to algal blooms, would necessitate additional treatment.

Phosphate fertilizer has been used extensively in forests in north Cumbria, but only on a limited scale in other parts of the county. Potassium has been used fairly widely in north Cumbria but there has been little use of nitrogen fertilizers in any Cumbrian forests to date. To obtain data on the impact of forest fertilizers on water quality it is necessary to look beyond Cumbria. Ground rock phosphate was applied to the Coal Burn catchment from the air some nine weeks prior to planting. Before fertilization, PO_4-P concentrations in the drainage waters ranged from 0.01 mg l^{-1} to 0.06 mg l^{-1}, with a mean of 0.016 mg l^{-1} (Robinson 1980). Immediately after application, the concentrations rose to 0.27 mg l^{-1}, probably as a result of rock phosphate falling directly into the drainage channel. Between storms, phosphate levels remained at 0.25–0.35 mg l^{-1}, but reached 1.5 mg l^{-1} during storms. Concentrations increased again when drainage operations started, ranging between 0.7 mg l^{-1} and 2.2 mg l^{-1}, and again during drain cleaning. Fiedler and Richter (1981) reported a similar rapid response to aerial application of fertilizers: peak values of 2.6 mg $PO_4-P\ l^{-1}$ were recorded after heavy rain. In a lysimeter study in southern Scotland, Malcolm and Cuttle (1983) found that phosphorus losses began 24 weeks after hand application of fertilizer, and were continuing, with little apparent decline, some three years after application. Harriman (1978) also reported that phosphate was still being lost to streams some three and a half years after aerial application to pole-stage Sitka

spruce: measured concentrations reached about 0.3 mg P l^{-1} shortly after fertilization and up to 0.1 mg l^{-1} three years later. He also found that nitrate levels doubled, to 0.4–0.5 mg N l^{-1}, after fertilizing with urea, declining to pre-treatment levels after about three years. The same study found that potassium concentrations declined to pre-treatment levels two years after applying potassium chloride.

It is clear from the above studies that fertilizers do find their way into drainage waters and may, in the case of phosphorus, influence water chemistry for up to three years. Although leaching of phosphorus occurs, there have as yet been no reported occurrences of algal blooms in reservoirs which could be directly linked to forest fertilization. Thus, it is likely that the worst fears of the water industry will not be realized. In many cases, the removal of phosphate by organisms in the stream channel and downstream dilution probably reduce concentrations to acceptable levels before they reach upland lakes or reservoirs. The biggest danger would seem to lie with small upland water supply reservoirs; however, careful management of phosphate fertilizer applications in these catchments can almost certainly overcome any potential problems.

Use of fertilizers is uncommon in forests and woodlands on the more fertile soils of the lowlands. Inputs of nitrogen, phosphorus and potassium to lowland waters from agriculture and sewage effluents will also be much greater than any possible contribution from forested land or woodland.

11.4 Herbicides and pesticides

Several of the more widely used pesticides are toxic substances which may have adverse effects on freshwater biota and, at higher levels, on human health. The chemicals can also produce an unpleasant taste in drinking water at levels much below those which would cause a risk to health. The European Community regulations covering drinking waters set a limit on total pesticide concentrations of 0.5 µg l^{-1} and for any individual pesticide of 0.1 µg l^{-1}. Pesticides are rarely used in Cumbrian forests, but they are used in other parts of Britain in response to specific pest outbreaks on conifers. Aerial application of pesticides would increase the risk of contamination of streams, as the chemicals can fall directly into surface waters.

Herbicides are used in Cumbria to suppress vegetation which is competing with young trees. They are applied from the ground as liquids using knapsack sprays or wick applicators, or as granules using pepperpot type applicators. The risk of contamination of watercourses is small, provided that published guidelines are followed (Sale, Tabbush & Lane 1983). In dry conditions, and with freely drained soils, most of the herbicide will be absorbed in the vegetation or soil. In wet conditions and poorly drained soils, surface flows could carry the chemicals into watercourses. It is suggested, therefore, that spraying be carried out in periods of dry, settled weather, particular care being taken in areas of poorly drained soils. In water catchment areas, it is suggested that, when using 2,4–D, the area treated at any one time may have to be restricted to allow for the effects of expected rain and surface runoff.

11.5 Harvesting

Clearfelling is the normal method of harvesting coniferous plantations in Britain. There have been many overseas studies of the impact of clearfelling on water quality, but none have been completed in Britain.

The impact of clearfelling is currently being investigated in Kershope Forest on the Bewcastle Fells (Pyatt *et al.* 1985; Adamson *et al.* 1987). The site lies at 225 m and is dominated by stagnohumic gley soils developed in clay-rich glacial till derived from the underlying Carboniferous rocks. The joint study, between the Institute of Terrestrial Ecology and the Forestry Commission, uses four two ha plots, three of which have now been felled. Each plot is isolated by deep drainage ditches and the drainage water from each is gauged using V-notch weirs. Drainage water is sampled at weekly intervals for chemical analysis. Data from this site show marked increases in the concentrations of nitrate, ammonium and potassium following felling (Figure 1), but a reduction in the concentration of most other solutes. Because of the large increases in water output following felling, the total outputs of most solutes increased, nitrate, ammonium and potassium showing the most dramatic rises. These changes in solute chemistry and solute outputs can be seen as a consequence of one or more of the modifications to the forest ecosystem which take place at felling.

Interception and transpiration are reduced as a result of canopy removal, leading to an increased flux of water through and out of the system. The removal of the canopy also reduces the 'capture' of particulate and gaseous material from the atmosphere. In the absence

of ground vegetation, root uptake ceases following felling. Changes in microclimate at ground level may produce changes in decomposition rates, and the resultant release of elements from organic materials. The lop and top left on the site produce a sudden large input of material and a rapid release of some of the elements they contain. The increased nitrate and ammonium concentrations and fluxes probably result from the reduction in root uptake plus the increased rates of decomposition of needles already on the ground before felling.

Reductions in the concentrations of sulphate, chloride, and sodium will result from a combination of reduced 'capture' from the atmosphere plus dilution due to the increased throughput of water.

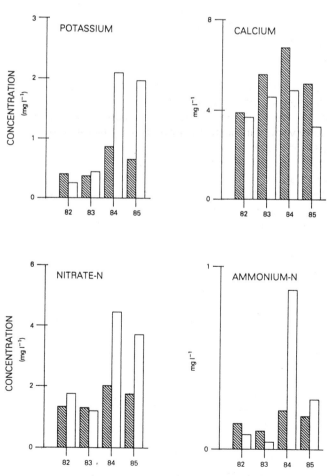

Figure 1. Annual mean concentration (mg l⁻¹) of four ions in drainage water, Kershope Forest, 1982–85. The hatched bars represent the drainage water from a control plot and the open bars an adjacent experimental plot. Felling took place on the experimental plot in 1983 (Adamson et al. 1987)

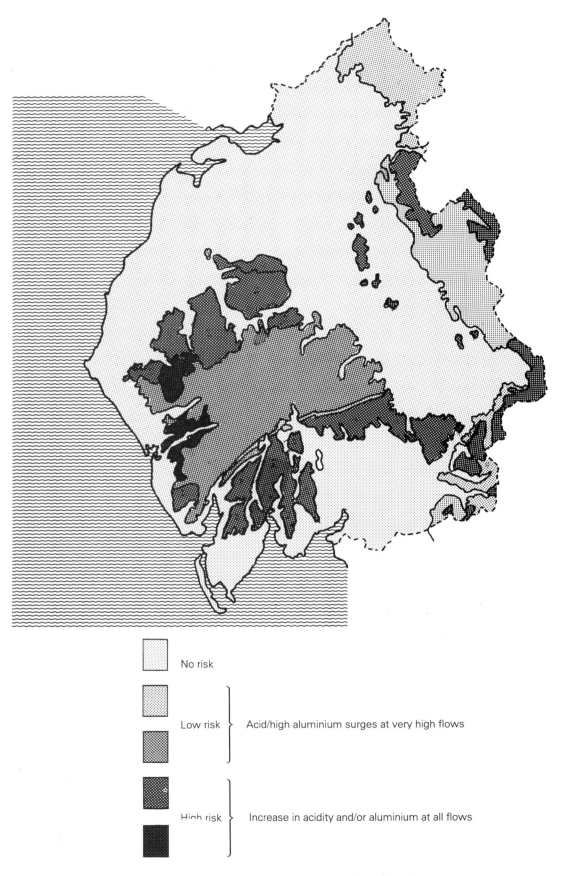

Figure 2. The relative risk of adverse impacts on water chemistry following afforestation

The increased solute concentrations are unlikely to have a significant effect on water supplies. Although maximum measured nitrate concentrations were close to the World Health Organisation's limits for drinking waters, these would be rapidly diluted downstream if felling coupes are not excessively large. Phosphate only leached into drainage waters during flood events, and again dilution would soon take place. Although the increased nutrient concentrations in drainage waters are unlikely to cause water treatment problems, they may lead to a undesirable increase in biological production immediately downstream of the felling.

11.6   Forestry and acid, high-aluminium waters

In recent years there has been much debate on the significance of atmospherically transported industrial pollutants. These pollutants have been linked to increased stream acidity and leaching of aluminium into streams (Underwood, Donald & Stoner 1987). These findings have, in turn, been linked to changes in the invertebrate populations of streams and reductions in fish populations (Crawshaw 1986). Several studies, from various parts of Britain, suggest that coniferous plantations can exacerbate the problem. Harriman and Morrison (1982), Stoner, Gee and Wade (1984), Reynolds *et al.* (1986) and Bull and Hall (1986) found higher acidity and greater concentrations of aluminium in streams draining established coniferous forest, than in adjacent streams draining unplanted moorland, on similar soils and geology. The apparent link between afforestation and increased acidity and aluminium concentrations is not found at all upland sites. It is most marked where acid soils overlie massive, base-poor bedrock. Here, acidity and aluminium concentrations are greater in forest streams at all levels of flow. The difference between forest and moorland streams will,

however, be greatest during high stream flows. Where acid soils overlie non-acid drift and/or bedrock, there is little difference at low flows between forest and moorland streams. In these conditions, water which has been in contact with the neutralizing drift or bedrock predominates. Only during floods will acid water penetrate downstream from the forest, because a significant proportion of the streamwater will have passed into the stream without making contact with the drift or bedrock.

Figure 2 combines information on the occurrence of acid soils in Cumbria with a ranking of the relative buffering or neutralizing capacity of bedrock, in an attempt to identify those areas where afforestation may be expected to have adverse effects on streamwater chemistry. The areas dominated by acid soils are taken from the 1:250 000 scale soil map of northern England (Jarvis *et al.* 1984). The soil associations, and soil types, grouped as acid are Winterhill (peats), Wilcocks (stagnohumic gley soils), Anglezarke and Crannymoor (humo-ferric podzols), and Skiddaw (humic rankers). All these soils are naturally acid, and the podzols and stagnohumic gley soils have exchange complexes dominated by aluminium.

The acid soil areas are subdivided into four classes on the basis of the buffering capacity of the underlying bedrock, using a grouping of rock types proposed by Kinneburgh and Edmunds (1984) (Table 1). These authors classified the map units on the 1:635 000 scale geology maps of the UK into four classes according to the sensitivity of groundwaters to acidification. This classification can be used as an index of the buffering capacity of the various rock types. Rocks with groundwaters most susceptible to acidification (class 1) will have the lowest buffering capacity; those with waters not at risk to acidification (class 5) will have the highest

Table 1. Categories adopted for classification of the solid geology map (1:625 000) of the UK, according to sensitivity to acidification (Kinneburgh & Edmunds 1984)

Category	Buffer capacity and/or impact on groundwaters	Rock type
1	Most areas susceptible to acidification, little or no buffer capacity, except where significant glacial drift	Granite and acid igneous rock, most metasediments, grits, quartz sandstones and decalcified sandstones, some Quaternary sands/drift
2	Many areas could be susceptible to acidification. Some buffer capacity due to traces of carbonate and mineral veining	Intermediate igneous rocks, metasediments free of carbonates, impure sandstones and shales, coal measure
3	Little general likelihood of acid susceptibility, except very locally	Basic and ultrabasic igneous rocks, calcareous sandstones, most drift and beach deposits, mudstones and marlstones
4	No likelihood of susceptibility. Infinite buffering capacity	Limestones, chalk, domomitic limestones and sediments

Farm woodlands

B D Everett
Middle Wood Centre, Wray, Lancaster

12.1 Introduction

Many farm woodlands in Cumbria are in a neglected state, offering few benefits to the farmer, and as a result their long-term survival is threatened. There are many existing benefits from well-managed woodlands on farms, and new uses for woodland produce are being investigated. The current problems of farm surpluses, decreasing farm incomes and loss of rural employment prompt the re-evaluation of farm woodlands.

12.2 Neglected woodlands

Neglect of woodland takes many forms. Woods may be left open to grazing stock to provide them with additional forage or temporary shelter. Such use removes the understorey shrubs, prevents tree regeneration and may lead to tree death through de-barking. Woods which were coppiced in the past may now have been left for 30 to 80 years without management, because of the decline in the market for coppice-wood products, resulting in a wood of tall, overcrowded trunks of similar age, with insufficient light reaching the ground for regeneration or the growth of shrubs and ground vegetation, and making it a poor wood for wildlife and game. These woods often appear as attractive deciduous areas from a distance, giving the impression that all is well within, although they are of poor structure and of little value for timber. They could be considered moribund, and are unlikely to recover without fairly harsh management.

The reason for the neglect may well be the farmer's lack of knowledge of woodland management, a consequence of farms becoming more and more specialized. This situation was highlighted by the Dartington Amenity Research Trust (1983), in a survey which revealed that only one per cent of farmers and landowners believed they had sufficient basic knowledge to manage their woods effectively. In order to encourage farmers to manage their own woodland they must be clearly shown the benefits that good management can offer to a farm.

12.3 Benefits of good management

Benefits to the farmer

Shelter
Many woods were originally planted to provide shelter for adjacent agricultural land. Livestock benefits by protection from blizzards, rain, cold winds or hot sun, especially at critical times such as lambing or shearing.

Shelter raises the ground temperature and allows the early growth of arable crops and grazing grass. It also reduces the mechanical damage to crops, such as lodging (blowing over) of corn or loss of blossom. On cultivated land, there will be a reduction of soil erosion by the wind. Where farm buildings are in the lea of a shelterbelt, there will be a reduction in heat loss and less water damage to the structure.

Game cover
Well-structured woods around the farm may provide valuable income for the farmer. H Edwards, a farmer in Norfolk, estimated that, by planting 5% of his farm with woodlands, he received a sporting rent of £12 ha^{-1} over the whole of his farm. Many Cumbrian woodlands have survived because of their importance for shooting.

Timber
Timber may be extracted for use on the farm, after conversion to fence posts, rails, gates or farm buildings, or for sale. At Middle Wood, we are using home-grown timber for new farm buildings, sawn on site using a chain mill. Oak fencing posts are riven.

Fuel
Most farmers extract firewood from their own woods, helping to reduce energy bills in the farmhouse.

Biological control
The presence of a diversity of wildlife in farm woods helps to prevent pests building up to numbers which would devastate a crop. The insectivorous birds are of particular value.

Benefits to the community

Woodlands can screen eyesores and help to maintain the appearance of the landscape, which is particularly important in the Lake District where tourism is a vital part of the local economy. It may even be important to the farmer in gaining additional income by providing overnight accommodation for tourists. Larger woods may be open to public access, providing important amenity areas. Woodlands are particularly valuable as they have a high capacity for holding people, without the feeling of crowding.

Woodlands provide a range of habitats for wildlife. Many people enjoy seeing wildlife around the farm; there is always a thrill in hearing woodpeckers (*Picus viridis*) 'yaffling' or seeing hares (*Lepus capensis*) playing in the early morning. Additionally, wild species represent a genetic bank of material that may be of use to future farmers or foresters in breeding domesticated varieties, eg breeding fast-growing birch (*Betula* spp.) or oak (*Quercus* spp.) with sweet acorns for human consumption.

12.4 Farm woodland management

Woodlands may have been neglected for a number of reasons, particularly the scale of operations and the financial implications for the farmer. Small woodlands of less than eight hectares will not produce commercially viable amounts of timber for sale, and woodland management and extraction cost money. Many small farms cannot afford to invest in the necessary equipment so the woodlands are ignored; they are regarded as a headache rather than an asset. The formation of farmers' co-operatives to purchase equipment for woodland management and to produce regular, commercially viable loads of timber will be essential to the future of small woods. Ideally, these co-operatives would be linked with an Agricultural Training Board, which could set up training workshops on the basic skills needed, eg the sharpening and care of chainsaws and circular saws and the identification of tree species. Many farmers would like to manage their neglected woodlands but do not know how; they need advice that takes into account the running of the remainder of the farm.

In recent years, there has been a gradual reduction in farmers' incomes and a sharp decline in rural employment prospects because of excess agricultural production in the European Community. In the UK, the domestic production of timber accounts for barely 10% of total requirements. The National Farmers' Union (1984) and the Centre for Agricultural Strategy (Carruthers & Jones 1983) have promoted the re-evaluation of farm woodlands. The establishment of farm woodland has become more attractive because of the grants available from the Agricultural Development and Advisory Service and the Forestry Commission to encourage the planting of broadleaved trees. The Minister for Agriculture has stated that forestry offers the most promising alternative use for land released from agriculture (Farmers Guardian 28/3/86), and it is possible that one million hectares of land will be planted with trees in the next 20 years.

From the farmer's point of view, there needs to be more research into markets and new uses for timber and other tree products. Farmers should be encouraged to plant unproductive areas of land, eg wet areas with problems of liver fluke or steep slopes where

erosion is common. Trees could be managed as dense coppice, providing fast-growing biomass to be used for energy, for example to be made into fuel blocks or chipped into large standard chips to fuel farm or industrial boilers. Alternatively, finer chips could be composted to provide organic fertilizers in the long term and heat during the breakdown period. There is also a large-scale potential for gasification and fermentation.

In combination with coppice as an energy crop, the value of the brushwood as a livestock feed should be considered. In Russia, brushwood is converted into a commercial animal food called Muka. The crude protein in tree leaves varies between 15% and 20%, while most proprietary sheep feeds range from 12% to 13%. Leaf silage may offer a valuable feed stuff to farmers, and will not depend on weather conditions.

The ability of farmers to produce low-cost buildings from their home-grown timber can help them adapt to changes in agricultural policy. The techniques being developed by the School of Woodland Industries (Savage 1986), where coppice timber or early conifer thinnings are used under tension, is worthy of attention.

With the development of coppice woodland, care must be taken not to go in the direction of monocultures which are of no value to wildlife in environmentally sensitive areas such as the Lake District, or crops which require high fertilizer and machinery inputs. It is important to produce a diverse woodland, using some native species and relying on the natural fertility of the land.

There is also potential on the farm for widely spaced trees, as in an orchard, to provide shelter for livestock and yet retain adequate grazing beneath. Lawson and Callaghan (see page 73) discuss the full potential for agroforestry. At Middle Wood we are hoping to plant alder (*Alnus* spp.) over an area of fell to provide shelter. When well established, the trees will be pollarded to provide additional feed for the sheep and also firewood.

12.5 Labour

The creation of job opportunities with the Manpower Services Commission (MSC), such as the recent Farm and Countryside Initiative, offers a new possibility for managing small woodlands on farms. The East Sussex Small Woodlands Project employs a full-time woodland advisor, and uses MSC teams of workers to manage uneconomic neglected woods and contractors or the farm's own labour to manage economically viable woods. This scheme appears to be successful in stimulating farmers to care for their woods.

Another possibility is the development of community woods (Woodland Trust 1986), where a local community takes on the responsibility for managing a wood and carries out all the work, but the farmer can sell the extracted timber. Such a development stimulates a positive approach to the woodlands from both the community and the farmer. The idea has been extended further by the Greenwood Trust, which manages the woods of Telford Development Corporation and involves local schools and colleges in converting the timber into furniture, sculptures and other interesting products.

12.6 Conclusion

Farmers can benefit considerably from managing existing woodlands and planting up new areas with trees. This benefit can only be achieved with a large input of information and training in new techniques, as well as an understanding of the existing advantages. The establishment of regional woodland co-ordinators would be of great assistance, but new markets for timbers must be researched so that woodlands will eventually provide income and employment for farmers. This development will help to stabilize the rural community, and will provide an attractive and diverse landscape with benefits to the scenery and increased environmental stability.

References

Carruthers, S.P. & Jones, M.R. 1983. *Biofuel production strategies for UK agriculture.* (CAS paper 13.) Reading: Centre for Agricultural Strategy.

Dartington Amenity Research Trust. 1983. *Small woods on farms.* (CCP 143.) Cheltenham: Countryside Commission.

National Farmers' Union. 1984. *The way forward: new directions for agricultural policy.* London: NFU.

Savage, D. 1986. The gleaning of the forests. *Woodworker*, March, 236–239.

Woodland Trust. 1986. *Community woodland resource pack.* Grantham: Woodland Trust.

Mixtures and mycorrhizas: the manipulation of nutrient cycling in forestry

A H F Brown and J Dighton
Institute of Terrestrial Ecology, Grange-over-Sands

13.1 Introduction

Although, in agriculture, soil fertility is primarily maintained by frequent fertilizer applications, this is not usually the case with forestry. The nutrients required by forest trees are normally provided by nutrient cycling, which is the circulation of nutrients in the forest ecosystem. It is only where this supply is seriously inhibited that repeated fertilizer applications are required. The increasing cost of fertilizers and their potentially environmentally undesirable effects, described by Hornung and Adamson (see page 55), may make their use in forestry even less attractive in future. Thus, it is important to see if we can, where necessary, boost fertility levels by enhancing nutrient cycling instead.

This paper describes two aspects of research which are based in Cumbria, at the Merlewood Research Station of the Institute of Terrestrial Ecology (ITE), but which have a very wide application. Historically, much of Merlewood's research has centred on the cycling of nutrients through woodland ecosystems. In the present paper, we aim to show the practical relevance of some of this work, by indicating how the forest manager may be able to manipulate nutrient cycling to improve plantation production.

A high proportion of the nutrients taken up by trees end up in the foliage, where they service photosynthesis and other physiological activities. Following leaf-fall, cycling of the nutrients remaining in the leaf litter depends on two main processes:

i. the release of nutrients by the decomposer activities of various organisms in the soil and litter (of trees and other plants): the adjacent planting of two different species of trees (tree mixtures) appears to enhance these activities.

ii. the ability of the trees to locate and take up these nutrients: manipulation of mycorrhizas may enhance this process. Mycorrhizas are an intimate symbiotic relationship between plant roots and fungi which are common on the majority of higher plants throughout the world.

13.2 Tree mixtures

Although there has been a recent renewal of interest in the possible benefits of tree mixtures, the idea that one species benefits from the presence of an admixed 'nurse' species is an old one. What sorts of mixtures lead to these benefits? Do all mixtures have this effect? What processes and mechanisms are involved?

The joint ITE/Forestry Commission mixtures and monocultures experiment at Gisburn (Forestry Commission's Bowland Forest in north-west England) provides some answers to these questions. Established in 1955, it contains four tree species – Scots pine (*Pinus sylvestris*), Norway spruce (*Picea abies*), sessile oak (*Quercus petraea*) and alder (*Alnus glutinosa*) – planted both as monocultures and as all possible two-species mixtures. The resulting ten treatments are replicated three times in 0.2 ha plots (Brown & Harrison 1983). The site has never been fertilized.

Height measurements of each species, repeated at intervals since planting, show that several different sorts of mixture effect occur; not all combinations are beneficial in terms of overall growth of the mixture. Data for the most recent measurements (age 26 years) are given in Table 1 (see also Lines 1982).

Table 1. Dominant heights (m) of four tree species at 26 years when grown pure and mixed: Gisburn, 1981. Heights in italics are for pure stands. Data for a given measured species (ie within columns) with different suffix letters (a, b, c, d) are significantly different at P <0.001, except for differences between spruce-with-alder and spruce-with-pine in which P <0.01

In mixture with	Measured species			
	Spruce	Oak	Alder	Pine
Spruce	*8.80*[a]	5.67[a]	7.57[a]	11.54
Oak	8.76[a]	*6.58*[b]	7.72[a]	11.34
Alder	9.84[b]	7.29[c]	*8.24*[b]	11.12
Pine	10.62[c]	8.82[d]	9.31[c]	*11.12*

NOTE: Some of these *plot* differences, although statistically significant as indicated, cannot strictly be ascribed to *treatment* effects because of the small number of replicates (blocks). Only differences associated with the beneficial effect of pine are rigorously referable to a treatment effect

Admixed pine stimulates height growth of all three other species without detriment to its own performance, whilst alder, although enhancing growth of oak and spruce, does so only at the expense of its own height growth. When oak and spruce are in mixture together, each grows worse than when grown separately.

Studies at Gisburn into the mechanisms involved in the mixture effects have been confined to the influence of admixed species on spruce. To identify whether improved height growth of spruce in the mixed stands was caused by improved nutrition, foliar analysis of the spruce, in pure and mixed stands, was carried out for nitrogen (N), phosphorus (P) and potassium (K), the three nutrients most commonly in short supply in forests. Results correspond with findings from other sites (O'Carroll 1978; McIntosh & Tabbush 1981; Taylor 1985), namely that N nutrition is a significant factor in the mixture effect. The improved height growth of spruce, in mixture with pine and alder, was associated with the highest levels of foliar N; the poorer growth of the pure spruce and of spruce admixed with oak was reflected in lower N concentrations (Figure 1). In addition, our results indicate that the bet-

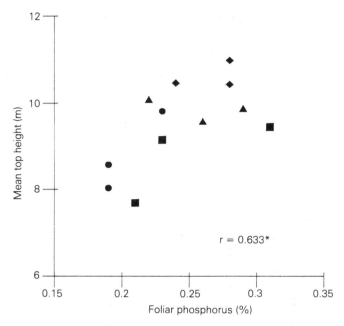

Figure 2. Relationships between tree heights and foliar phosphorus (means per plot for each of three blocks): Gisburn, 1981. Symbols as Figure 1

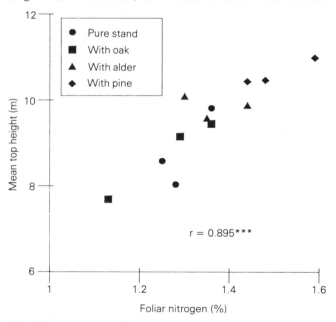

Figure 1. Relationship between tree heights and foliar nitrogen (means per plot for each of three blocks): Gisburn, 1981

ter growth is also associated with higher foliar P (Figure 2). On the other hand, there was no relationship between foliar K and spruce heights.

In the case of the spruce trees which benefit from the presence of pine and alder, where does the extra N and P come from? The fact that the nutritional benefit appears to be confined to these two elements suggests an organic matter source. As the rate of breakdown of such material is known to limit cycling of these two nutrients more than any others, an estimate

Figure 3. Layout of a mixed plot at Gisburn. Away from the edge of the plot each species is arranged in alternating blocks of three trees by six trees

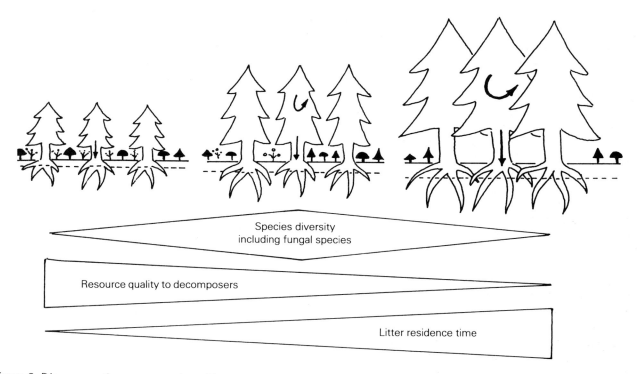

Figure 9. Diagrammatic representation of forest succession showing changes in ground flora and mycorrhizal fungi related to changes in resources entering the decomposer community (Dighton & Mason 1985)

by Went and Stark (1968), based on observations in tropical forests. There is evidence from our work and in the literature that certain mycorrhizal fungi, particularly under stressed environmental conditions, are able to act as decomposers. Dighton (1983) compared the phosphatase and phytase enzyme activity of known mycorrhizal and saprotrophic fungi, and found that some of the mycorrhizal species were capable of producing larger amounts of these enzymes than the saprotrophic fungi (Table 3). This finding implies a potential of these fungi to degrade complex organic phosphates. Similar studies have been made (Ho & Zak 1979; Alexander & Hardy 1981) to show phosphatase activity in other mycorrhizal fungi. Similarly, Giltrap (1982) has shown that polyphenol oxidases are produced by the mycorrhizal fungus *Lactarius* spp.; Norkrans (1950), Oelbe (1982) and Linkins and Antibus (1981) have shown cellulase activity of *Tricholoma* and willow (*Salix* spp.) mycorrhizas.

For selection of potentially suitable mycorrhizal species/strains to enhance tree crop yield, a number of factors must be considered, many of which depend on

Table 3. Acid phosphatase production by fungal mycelia in Hagem's medium with 10 ppm orthophosphate-P or inositol hexaphosphate-P after 42 days' growth at 20 °C (Dighton 1983)

Fungus	Orthophosphate-P	Inositol hexaphosphate-P
Hebeloma crustuliniforme	29.8	23.9
Lactarius rufus	26.3	22.2
Paxillus involutus	17.1	10.1
Lactarius pubescens	12.8	7.6
Amanita muscaria	5.8	8.8
Suillus luteus	5.0	9.2
Marasmius androsaceus (s)	20.0	1.7
Mycena galopus (s)	0.9	0.2

(s) = saprotrophic fungus, the remainder are mycorrhizal

soil properties. From what stage of the succession should the fungi be selected? Have they been proven to enhance growth/nutrient content of the tree species under consideration? Are they able to compete against indigenous mycorrhizas or are they rapidly replaced?

Will the fungus survive in the soil type in which it is to be placed? If the soil is highly organic, containing recalcitrant litters, has the fungus the enzyme potential to act as a decomposer? What is the potential outcome of competition between the mycorrhizal fungus and indigenous saprotrophic fungi? Obviously, we are only just beginning to unravel some of the answers to these complex questions.

Nevertheless, present knowledge of the principles involved has now been applied to field trials near Hexham and near Jedburgh. Forest transplants were inoculated with selected mycorrhizal fungi prior to planting. Although the trials are only three to four years old, preliminary results for certain of the mycorrhizal selections look very promising, showing enhanced growth rates compared with the control transplants. The latter are likely to have mycorrhizas appropriate to a nutrient-rich mineral nursery soil, which is unlikely to be found in most field situations. It is now important for us to examine the potentially useful fungi to discover their physiological attributes which make them so successful.

13.4   Conclusion

Research into both tree mixtures and mycorrhizal inoculation is continuing at Merlewood and other research establishments. In the case of tree mixtures, the purpose is to understand fully the soil processes responsible for the success of this existing management practice, so that its full potential may be realized. In the case of mycorrhizal inoculation, the mechanisms by which it would operate have been defined from the outset, and research is directed towards developing a practical management technique. Both research areas have the common aim of enhancing woodland growth by manipulating nutrient cycles.

References

Alexander, I.J. & Hardy, K. 1981. Surface phosphatase activity of Sitka spruce mycorrhizas from a serpentine site. *Soil Biology & Biochemistry*, **13**, 301–305.

Brown, A.H.F. & Harrison, A.F. 1983. Effects of tree mixtures on earthworm populations and nitrogen and phosphorus status in Norway spruce (*Picea abies*) stands. In: *New trends in soil biology*, edited by P. Lebrun and others, 101–108. (Proc. 8th International Colloquium of Soil Zoology.) Ottignies-Louvain-la-Neuve: Diew-Brichart.

Brown, A.H.F. & Howson, G. 1988. Changes in tensile strength loss of cotton strips with season and depth under 4 tree species. In:

Cotton strip assay: an index of decomposition in soil, edited by A.F. Harrison, P.M. Latter & D.W.H. Walton, 86–89. (ITE symposium no. 24.) Grange-over-Sands: Institute of Terrestrial Ecology.

Chu-Chou, M. 1979. Mycorrhizal fungi of *Pinus radiata* in New Zealand. *Soil Biology & Biochemistry*, **11**, 557–562.

Chu-Chou, M. & Grace, L.J. 1981. Mycorrhizal fungi of *Pseudotsuga menziesii* in the north island of New Zealand. *Soil Biology & Biochemistry*, **13**, 247–249.

Dighton, J. 1983. Phosphatase production by mycorrhizal fungi. *Plant and Soil*, **71**, 455–462.

Dighton, J. & Mason, P.A. 1985. Mycorrhizal dynamics during forest tree development. In: *Developmental biology of higher fungi*, edited by D. Moore, L.A. Casselton, D.A. Wood & J.C. Frankland, 117–139. Cambridge: Cambridge University Press.

Dighton, J., Poskitt, J.M. & Howard, D.M. 1986. Changes in occurrence of basidiomycete fruit bodies during forest stand development with specific reference to mycorrhizal species. *Transactions of the British Mycological Society*, **87**, 163–171.

Giltrap, N.J. 1982. Production of polyphenol oxidases by ectomycorrhizal fungi with special reference to *Lactarius* spp. *Transactions of the British Mycological Society*, **78**, 75–81.

Ho, I. & Zak, J.C. 1979. Acid phosphatase activity of six ectomycorrhizal fungi. *Canadian Journal of Botany*, **57**, 1203–1205.

Latter, P.M. & Howson, G. 1977. The use of cotton strips to indicate cellulose decomposition in the field. *Pedobiologia*, **17**, 145–155.

Lines, R. 1982. Species: mixture experiments. *Report of Forest Research, 1982*, 13–14. Edinburgh: Forestry Commission.

Linkins, A.E. & Antibus, R.K. 1981. Mycorrhizae of *Salix rotundifolia* in coastal arctic tundra. In: *Arctic and alpine mycology*, edited by G.A. Laursen & J.F. Ammirati, 509–305. Washington: Washington University Press.

McIntosh, R. & Tabbush, M. 1981. Nutrition. *Report of Forest Research, 1981*, 21–22. Edinburgh: Forestry Commission.

Mason, P.A., Last, F.T., Pelham, J. & Ingleby, K. 1982. Ecology of some fungi associated with an ageing stand of birches (*Betula pendula* and *Betula pubescens*). *Forest Ecology and Management*, **4**, 19–37.

Mosse, B., Stribley, D.P. & Le Tacom, F. 1981. Ecology of mycorrhizae and mycorrhizal fungi. *Advances in Microbial Ecology*, **5**, 137–210.

Norkrans, B. 1950. Studies in growth and cellulytic enzymes of *Tricholoma*. *Symbolae Botanicae Upsaliensis*, **11**, 1–126.

O'Carroll, N. 1978. The nursing of Sitka spruce. 1. Japanese larch. *Irish Forestry*, **35**, 60–65.

Oelbe, M. 1982. *Untersuchungen uber einige Kohlenhydratabbauende Enzyme des Mykorrhizapilzes* Tricholoma aurantium. MSc thesis, Georg-August-University, Gottingen.

Taylor, C.M.A. 1985. The return of nursing mixtures. *Forestry and British Timber*, **14 (5)**, 18–19.

Went, F.W. & Stark, N. 1968. The biological and mechanical role of soil fungi. *Proceedings of National Academy of Science, USA*, **60**, 497–504.

Agroforestry

G J Lawson and T V Callaghan
Institute of Terrestrial Ecology, Grange-over-Sands

14.1 Introduction

Agroforestry is a term which describes systems in which trees, animals and/or crops are grown together in intimate mixtures. The term does not include farm woodlands which do not involve significant biological or environmental interactions between the woodland and agricultural components.

A number of papers in this volume have discussed the possibility of increasing timber production in Cumbria, in response to Britain's present high rate of timber imports and excess agricultural production. Agroforestry could contribute to this increased timber production in a manner which would be attractive to the farming community because land would sustain a significant agricultural income whilst the trees were maturing.

There are several types of agroforestry. *Silvoarable systems* are mixtures of trees and crops, while *silvopastoralism* describes intimate mixtures of trees and animals. *Agrenforestry* is a term used here to emphasize the co-production of bioenergy crops with agricultural and timber crops. Agrenforestry, described later in more detail, could involve strips of energy coppice amongst agricultural crops, or the use of coppice beneath wide-spaced standard trees.

It has been suggested (Lawson 1987) that there are five possible uses for rural land — food, fibre, fuel, pharmaceuticals and fun. This paper moves up the alphabet to discuss five criteria for land use decisions: economics, energy, environment, employment and enjoyment.

14.2 Economics

Agroforestry is an unusual land use in present-day Britain. However, there is a long history of coppicing and pollarding which Satchell has described for Cumbria (see page 6). Multiple use of trees has an even longer history than coppicing, and Satchell has discussed the argument that excessive lopping for fodder caused the extensive decline of elm (*Ulmus* spp.) in pollen diagrams, around 3000 BC. The starch and protein content of leaves from several tree species can exceed that in good grass (Russell 1947). Medieval parkland contained herds of deer, cattle and swine grazing under widely spaced trees. Wooded commons made extensive use of pollarding, and winter grazing within forests remains vital to the survival of deer and sheep in some areas.

However, despite this long-standing experience of agroforestry, it is hard to predict its current profitability, which will be much influenced by factors such as whether:

i. high pruning of conifers can sustain yields

ii. a sufficient premium will develop for veneer-quality timber

iii. intensive fertilization will damage timber quality

iv. bark-stripping can be limited by good grazing management

v. the shelter provided for stock compensates for the grazing loss.

Despite the uncertainties, several economic models of agroforestry have been developed. They are based on rather heroic biological assumptions, but experimental evidence is beginning to verify the predictions.

New Zealand

New Zealand provides an example of agroforestry as it may apply in Britain, and several large-scale experiments have been continuing there for at least 12 years. Agroforestry in New Zealand was a consequence of suggested changes in the management of radiata pine (*Pinus radiata*) on good-quality sites. This 'direct sawlog regime' advocated a low number of crop trees, early thinning to waste instead of productive thinning, and intensive pruning of lower branches (Fenton & Sutton 1968). The possibility of grazing in these widely spaced forests was almost coincidental, but it has engendered considerable interest amongst New Zealand farmers. There are now at least 30 000 ha of agroforestry and 70 000 ha of forests with a grazing component (Percival & Hawke 1985).

The carrying capacity of hill pasture has been examined at different tree densities (Figure 1), and the profitability of agroforestry appears to compare well with conventional agriculture (Table 1). It should be noted that New Zealand now has none of the subsidies for agriculture or forestry which confuse land use comparisons in the European Community (EC). Predictions from all four sites in Table 1 suggest that agroforestry has an internal rate of return on investment in excess of 10%. However, further sensitivity analyses show that agroforestry would be uneconomic with higher stocking rates, infertile sites, poor silvicultural management, long haulage distances, high harvesting and sawmill costs, and a low stemwood price. The optimum spacing was found to be 100

Table 1.   Comparison of the profitability of agroforestry at four locations in New Zealand (assuming 100 stems ha^{-1}, 10% discount rate, January 1984 prices, average stocking units ha^{-1}, no premium for quality timber, average site index) (Arthur-Worsop 1985)

| | Net present value (NZ\$ ha^{-1}) | | | |
	Whatawhata (hard hill land)	Tikitere (easy hill land)	Invermay (rolling hills)	Akatore (steep, less fertile)
BENEFITS				
Gross forestry revenue	2524	2842	2343	2412
Forest grazing gross margin (GM)	1335	1646	1519	1406
Total	3859	4488	3862	3818
COSTS				
Forestry costs	1786	1622	1230	1498
Open grazing opportunity costs (GM)	1773	2428	2393	2185
Total	3559	4050	3623	3683
Net effect of agroforestry	+300	+438	+239	+135
Internal rate of return	11.6	11.8	11.0	10.6

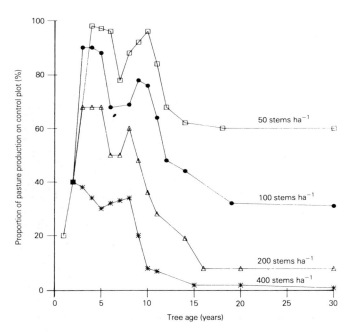

Figure 1. The measured (years 4–11) and predicted (years 12–30) effect of radiata pine on livestock numbers in New Zealand hill pasture (Percival & Hawke 1985)

stems ha^{-1}. As always with comparisons between forestry and agriculture, the assumption of a low discount rate will favour systems with a high revenue from trees at the end of a rotation, rather than agricultural systems with a guaranteed annual return.

Chile

A multidisciplinary group in southern Chile initiated a number of agroforestry experiments in 1977. As in New Zealand, their main interest has been in wide-spaced and pruned radiata pine, but experiments have also been established using southern beech (*Nothofagus* spp.), alder (*Alnus glutinosa*), chestnut (*Castanea sativa*), and Douglas fir (*Pseudotsuga menziesii*) (Penalosa, Herve & Sobarzo 1985). Sheepmeat production within six-year-old pines at 100 stems ha^{-1} has reached 250 kg ha^{-1}, compared to 214 kg ha^{-1} prior to planting. This increase is attributed to the conservation of soil moisture caused by shade from the tree canopy.

Italy

Several fertile valleys in Italy, particularly that of the River Po, demonstrate a system where poplars (*Populus* spp.) compose around 20% of the farming area. The trees are grown at wide spacing (8–10 m) in double or single rows, and are often concentrated at field boundaries or road verges. For the first five years of growth, soil cultivation between rows of poplar increases tree height and girth increment. These benefits remain during the whole rotation (Food and Agriculture Organisation 1980). Farm manure is applied routinely, and appears not to cause the wood quality problems which occur when conifers are heavily fertilized. Forage maize, wheat, pulses and root vegetables are all used in the early years of a plantation (Plate 5), to be followed by the grazing of cattle.

Plate 5. Casale Monferato, Italy. Two-year-old rooted cuttings of poplar (*Populus* × *euramericana*) underplanted with arable crops. (Photograph G J Lawson)

Plate 6. Bruton, Somerset. New parkland dominated by 80-year-old oak (*Quercus* spp.). Agricultural grants and firewood revenues covered the cost of conversion from neglected scrub. (Photograph G J Lawson)

Prevosto *et al.* (reported in Food and Agriculture Organization 1980) conducted extensive trials on the reduction in yield of different crops caused by wide-spaced rows of poplar. These yield reductions, together with the higher costs of cultivation, have been incorporated in an economic assessment of the implications of gradually introducing poplars on to a typical farm of the Po Valley. Prior to poplar planting, the farm yielded \$429 ha^{-1} (1975 prices), while at the end of a ten-year rotation the revenues per hectare had reached \$509, \$589 and \$669 for 10%, 20% and 30% poplar respectively at a 10% discount rate. Poplars may have an even greater advantage in areas where crop yields are reduced by wind exposure, or where the prunings are used for animal fodder. However, in recent years, increasing agricultural subsidies and a depressed market for timber in Italy have significantly reduced the prevalence of agroforestry on fertile farmland.

Britain

The Hill Farming Research Organisation and Forestry Commission have jointly considered the economics of a combined conifer and sheep silvopastoral system (Maxwell 1986; Sibbald *et al.* 1987). Data from a number of species are used to predict conifer growth, shading effect and timber yield at a range of planting densities, and under two sheep production systems (Table 2). Under the given assumptions, the model indicates that silvopastoralism may be the most profitable land use on upland farms, and predicts that a density of 100 stems ha^{-1} is preferable to 400 stems ha^{-1}.

Another modelling exercise was performed jointly by the Animal and Grassland Research Institute and the Forestry Commission, this time for lowland grassland and broadleaved trees. Simple competition models for light, moisture and nutrients were used to predict the yield of grass beneath ash (*Fraxinus excelsior*) trees at spacings of up to 200 ha^{-1}, receiving different levels of nitrogen fertilizer. High-quality timber production was predicted to be much more profitable than early felling for firewood, and 100 stems ha^{-1} was considered the optimum spacing. Again, the discount rate selected considerably influences profitability. At a 5% rate, and particularly with fertilizer levels less than 150 kg ha^{-1}, the model suggests significant advantages for hardwood silvopastoral systems in the lowlands (Doyle, Evans & Rossiter 1987).

Crop interaction economics

In agroforestry one can distinguish between *symbiotic, independent* and *competitive* relationships.

In a *symbiotic* relationship, the crops have positive influences on the growth of each other. There are many possible examples in UK agroforests: shelter may increase pasture yields by raising ground temperature in spring and relieving moisture stress in summer (Marshall 1967); cultivating the soil between rows of poplar during the early years of a rotation can increase the growth rate of the trees (de Damas 1978); nitrogen-fixing understories will enhance the growth of trees on poor soils (O'Carroll 1982).

An *independent* relationship exists if the two crops have no influence on each other, for example if they use labour at different times of year, or if the trees and herbaceous crops are using different pool of nutrients or moisture.

Competitive relationships exist when the two crops compete for resources of light, moisture, nutrients, labour, land or capital.

Table 2. Net present value of various silvopastoral options compared with conventional agriculture or forestry (£ ha^{-1}, 5% discount rate, assumed planting densities 100 and 400 stems ha^{-1}, Douglas fir, yield class 20, individual protection assumed for trees – allowing grazing from the first year, upland sheep = 10 greyface ewes ha^{-1}, hillsheep = 2.5 blackface ewes ha^{-1}, premium of 7% assumed for high timber quality, 4 prunings assumed) (Maxwell 1986)

		Hill			Upland		
	All forest	All agri-culture	Agro-forestry (100 ha^{-1})	Agro-forestry (400 ha^{-1})	All agri-culture	Agro-forestry (100 ha^{-1})	Agro-forestry (400 ha^{-1})
Forestry	1129	—	268	638	—	268	638
Agriculture	—	555	525	433	4234	4005	3304
Total	1129	555	793	1071	4234	4273	3942

These three types of interaction are expressed in Figure 2 at different points on the revenue curve be-

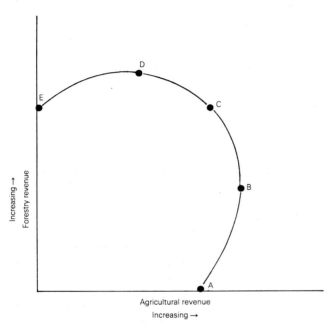

Figure 2. Revenue interactions for different mixtures of agriculture and forestry (adapted from Filius 1982, see text for discussion)

tween pure agriculture (A) and pure forestry (E). The line, A to E, represents increasing tree density. At very low tree stocking rates (A–B), the sheltering effect of trees may benefit agricultural production or control erosion. Timber revenue rises rapidly as density increases because there is little inter-tree competition. This is, therefore, a period of symbiotic interaction. The interaction becomes competitive as the tree density is further increased and agricultural production declines (B–D). Further increases in density reduce the net discounted revenue from forestry because of thinning costs or early harvesting, and agricultural revenue soon reaches zero (D–E). At points B and D, there is an independent relationship because a marginal increase in one crop makes no difference to revenue from the other. The optimum density is at point C, where a marginal decrease in forestry revenue is matched by an equal increase in agricultural revenue.

Note that a convex shape for this 'crop interaction curve' favours agroforestry. A concave curve would indicate that agroforestry was not economic, and

would occur if the two crops were competitive throughout the rotation, or if significant economies of scale exist in the monocultures.

Given enough biological information, Figure 2 is a useful economic nomograph to decide on the balance between two competing uses of a resource. When applied to the allocation of land between agriculture and forestry, it shows that the points of maximum agricultural revenue (B) and forestry revenue (D) need not maximize the use of the land resource. A useful extension of the method would be to apply shadow pricing to reflect the true value of each product to government and to account for the energetic (section 14.3), environmental (section 14.4) and social (section 14.5) implications.

Non-accountable factors

The economic models described above are overly simplistic, and this fact is emphasized by their authors. Interactions between trees and understorey vegetation or animals are very diverse (Appendix I), and demand much more research.

Wide-spaced 'silvopastoral shelter strips' in the uplands would have several economic advantages, some of which are rather difficult to quantify.

i. Shelter would be provided within the strip as well as outside it.

ii. Shelter extends further downwind from a sparse shelterbelt than from a dense one (Figure 3).

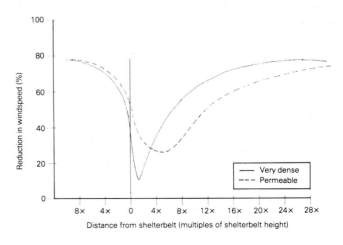

Figure 3. Differences in pattern of windspeed reduction between a permeable and a very dense shelterbelt (after Caborn 1965)

iii. For small areas, individual tree shelters are significantly cheaper than fencing. Wide-spaced planting will therefore be economic in smaller units, which will suit a varied topography.

iv. Grazing remains possible in silvopastoral strips (Figure 1), thereby reducing the cash flow problems incurred in plantations of conventional spacing.

v. Trees planted at wide spacing tend to develop spreading root systems, and will suffer less from windblow after thinning.

vi. Control of deer and foxes will be easier than in dense forest.

vii. Prunings from lower branches may be a useful food supplement for cattle and sheep.

There are also some disadvantages of agroforestry which are difficult to quantify. These include the effect of heavy fertilizer applications on timber quality, and the damage to trees caused by soil compaction or bark-stripping.

Agroforestry, therefore, can have implications for farmers which are not amenable to economic accounting. Farmers interested in forestry may plant tree cover for game birds, shelter for livestock or as a use for areas of their farm which are unsuitable for agriculture. Several mutual benefits were described by Mutch and Hutchinson (1980), where case studies of 13 upland farms indicated that afforestation of approximately 25% of the land was accompanied by an increase in livestock production of 33%, and an increase in employment of more than 50%.

Macro-economic factors

The micro-economic models also fail to account for the differences in subsidies given to farming and forestry. They are notoriously difficult to disentangle, but the following statistics emphasize that the balance of official support is highly likely to move towards woody crops, and away from intensive agriculture.

i. Obvious subsidies given to UK agriculture in 1985 were £2.21 billion, compared with a net farming income of only £1.15 billion (Ministry of Agriculture, Fisheries and Food 1986).

ii. Some estimates suggest that up to 2.4 Mha in the UK will be producing food surpluses by the year 2000 (Brown 1988). Even the Ministry of Agriculture, Fisheries and Food (MAFF) has accepted that land producing surpluses may exceed one Mha by the year 2000 (Anderson 1987). Comparable figures for the EC range up to 15 Mha.

iii. Of the timber used in the EC, 60% is imported. As long ago as 1959, the Mansholt Plan envisaged the need to transfer 5 Mha from farming to forestry. Imports of timber and timber products into the UK amount to £4.5 billion annually. The Forest Action Plan, currently under discussion in the EC, suggests many measures which would support the development of both agroforestry and plantation forestry, and would introduce support for wood marketing associations and wood utilization industries (Commission of European Communities 1986).

iv. In the EC, 45% of the fuel used is imported, and pressures are developing to subsidize the production of alcohols, or other biofuels, from energy plantations.

14.3 Bioenergy

In their summary of the potential for wood as fuel in the UK, the Department of Energy (Price & Mitchell 1985) predicts that 0.65 million tonnes coal equivalent (Mtce) could be raised from existing wood residues by the end of the century. If supplemented by wood energy plantations, this contribution could rise to 1.0 Mtce for industrial markets and 2.8 Mtce for the domestic sector. A doubling of energy prices would increase the availability of fuelwood three- to four-fold. The maximum predicted contribution at these increased prices would be around 13 Mtce, and this figure represents around 4% of total energy consumption in the UK, or 12% of the current consumption of coal. A study has been made of the land in Great Britain which could be available for energy forestry (Department of Energy 1987). This study suggested that up to 4.6 Mha could (at 1977 prices) be used profitably for energy coppicing, or for the enhanced used of residues from a modified form of single-stem forestry.

The major impediment to increased utilization of wood fuel is not the cost of production, but the fact that the bulk of wood residues is produced too far away from the heavily populated areas which sustain the best prices. There is considerable need for an economic comparison of the profitability of biomass plantations with conventional forestry and agriculture, where the conventional land uses have been stripped of the complex structure of planting grants, price support and tax relief.

Even without subsidies, energy coppices of fast-growing hardwoods like willow (*Salix* spp.) and poplar can be a profitable use of marginal agricultural land, provided that secure markets have been established (Scott *et al.* 1986). The best UK study of energy coppice comes from Northern Ireland, in conjunction with the Long Ashton Research Station (McElroy & Dawson 1986). Annual yields of 12–15 t ha^{-1} have been achieved over a nine-year rotation using a particular clone of willow (*Salix* 'Aquatica gigantea'). The experiments were conducted on surface mineral gley soils, which are marginal for agriculture. Three points are of particular note:

i. the costs of fertilizer additions were incompletely met by revenue from the resulting small increases in yield (up to 20%);

ii. the annual energy output from coppiced willow was 136 GJ ha^{-1}, compared with a net energy output from grass on comparable beef-producing land of 40 GJ ha^{-1};

iii. this variety of willow has recently been severely damaged by a rust fungus, illustrating the danger of an over-reliance on individual clones.

The Forestry Commission has also established energy coppice trials in different parts of the country ranging from the Cambridgeshire Fens to a gleyed site at 250 m OD in Scotland. Poplar and willow are confirmed as the most reliable producers, whilst establishment problems have been experienced with alder, and frost or disease problems are apparent for southern beech and *Eucalyptus* (Booth 1988).

Aberdeen University has established 11 trial single-stem plantations at 1 m × 1 m spacing in different parts of the country. Ten tree species have been planted, and the economic harvesting period should be around 20–30 years. Aberdeen University is also engaged in two large-scale (2 ha) coppice trials, with the intention of establishing 'production level' yields and management costs (Mitchell 1987).

Whilst very dense plantations offer the shortest rotations, they also maximize planting costs. Recent opinion, therefore, favours planting at lesser densities (3000–4000 stems ha^{-1}) for short-rotation forestry (Zsuffa & Barkley 1985). Others argue that, rather than establishing dedicated energy plantations, we should use more plantations at conventional spacing, and harvest the residues for energy. Prevosto (1979) compared the economics of conventional and close-spaced poplar, and concluded that spacings of 5–6 m were clearly more profitable because of the higher proportions of sawn wood and veneer wood which they contain, despite the fact that they produce less wood in

total and at longer rotations. This conclusion seems likely to pertain also in lowland Britain.

Where conditions are suitable for productive energy coppice, it is likely that greater profits would be made using suitable timber trees. Hardwoods such as southern beech, poplar, *Eucalyptus*, red alder (*Alnus rubra*), ash, cherry (*Prunus* spp.) and sycamore can grow extremely rapidly in suitable habitats. Many conifers, such as Douglas fir, will also perform well in lowland soils, and the economics of using good land to produce wood for burning must be questioned.

Fortunately, it is possible to combine energy and timber cropping on the same unit of land. Like most good ideas it is not new, and involves a rediscovery of the coppice-with-standards system. Selected specimens of light-demanding and open-canopied trees, like ash, southern beech or poplar, would be established at wide spacing and underplanted with shade-bearing trees like hazel (*Corylus avellana*) or shrubs like *Rhododendron* (Lawson 1987). These would be coppiced regularly, and lower branches from the timber trees would be high-pruned at the same time. Whilst the species mentioned and the intensity of management will not replace the environmental diversity of old coppice, many of the disadvantages of monocultures discussed in the next section will be avoided.

14.4 Environment

Species diversity

Agroforests are likely to be more diverse in structure and species than monocultures (Callaghan *et al.* 1986). This is true in the case of animals and plants, and it also applies to the less obvious microflora and microfauna. However, tree monocultures can have biologically rich phases, and young forest plantations, which are protected from grazing, will support a larger population of small mammals and predatory birds than will silvopastoral mixtures of Sitka spruce (*Picea sitchensis*) and rye-grass (*Lolium perenne*).

Species of open-ground birds will certainly be discouraged by any extensive tree planting. Golden plover (*Pluvialis apricaria*), red grouse (*Lagopus lagopus scoticus*), dunlin (*Calidris alpina*), snipe (*Gallinago gallinago*), curlew (*Numenius arquata*), meadow-pipit (*Anthus pratensis*), ringed plover (*Charadrius hiaticula*), lapwing (*Vanellus vanellus*), wheatear (*Oenanthe oenanthe*), stone curlew (*Burhinus oedicnemus*) and skylark

(*Alauda arvensis*) are likely to be in this category (Reed 1982). A larger list of birds benefit from pre-thicket stage plantations: these include the willow warbler (*Phylloscopus trochilis*), grasshopper warbler (*Locustella naevia*), chaffinch (*Fringilla coelebs*), whinchat (*Saxicola rubetra*), stonechat (*Saxicola torquata*), woodlark (*Lullula arborea*), tree pipit (*Anthus trivialis*), whitethroat (*Sylvia communis*) and black grouse (*Lyrurus tetrix*). Mammals also benefit considerably from the diversity within an open woodland (Staines 1986). It is unlikely that heavily grazed silvopastoral systems will provide great benefit to some of these species. Nevertheless, the open structure of an agroforest will duplicate many of the advantages of an immature woodland. For example, the nesting and brooding sites for many bird species are likely to be brought closer together. Certainly, the inhospitable conditions of a thicket-stage plantation will be avoided, and it is thought that silvopastoralism will be of net benefit to a wide variety of wildlife. The benefit of silvoarable systems is clearer, because not only will the two main crops be juxtaposed, but the uncultivated ground within rows of trees will provide an additional habitat.

In the absence of any experimental evidence about the wildlife implication of agroforestry, we must draw parallels from existing evidence in agriculture and forestry. Using bird populations as an indicator of habitat 'richness', it is clear that a greater diversity of agricultural crops increases wildlife interest (Figure 4a), and more diverse woodland structures cause similar increases in numbers of species (Figure 4b). Agroforestry should, therefore, aim towards diverse mixtures of crop types and structures.

Pests and pathogens

Species diversity will often lead to a greater resistance to disease and predation. The western spruce budworm (*Christoneura fumiferana*) causes severe damage to monocultures of balsam fir (*Abies balsamifera*) and Douglas fir, but has much less effect on mixtures (Fauss & Pierce 1969). Similarly, pine looper moth (*Bupalis piniaris*) has been reported to cause much less damage in mixed stands of oak (*Quercus* spp.) and Scots pine (*Pinus sylvestris*), than it does in pure pine stands. The pure stands are thought to have fewer parasites and other checks on the looper moth (Niemeyer 1986).

However, tree mixtures are not always healthier than monocultures. The spruce gall aphid (*Adelges cooleyi*)

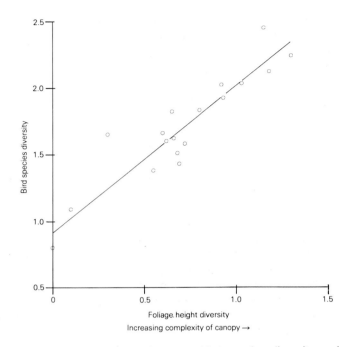

Figure 4a. Relationship between bird species diversity and foliage height diversity when examining a variety of mature woods with broadleaved or coniferous species or mixtures (Newton & Moss 1981)

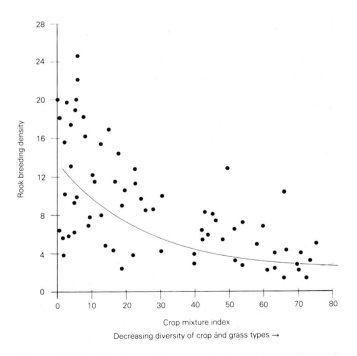

Figure 4b. Relationship between rook breeding density and an index of crop diversity (lower index values have higher diversity – Brenchley 1984)

alternates its life cycle between Douglas fir and spruce. The presence of both species can lead to significant damage to the spruce. Similarly, the rust fungus *Melampsora pinitorqua* alternates its life cycle between pine and aspen (*Populus tremula*), and causes significant damage to pines in mixed stands (Savill & Evans 1986).

The presence of lines of trees will make crop spraying more difficult. However, row spacings can be selected to match existing farm machinery; for example, a 14 m spacing permits conventional 12 m booms and 4 m combine harvesters to be used. Useful research is proceeding in MAFF's Boxworth project, and elsewhere, to quantify the environmental effects and costs of different levels of pesticide application (Hardy 1986). Less intensive management of cereals may be justified by the possibility that pesticides may cause further infestations by reducing the population of predatory bugs and mites (Chaboussou 1986). Hedgerows and field boundaries are the overwintering sites for these predatory insects, and similar habitats are available in agroforests. Seed-eating birds will thrive in close mixtures of trees and crops, but it is likely that cereals will be grown only during the first four to five years of a silvoarable rotation. Insectivorous birds will also benefit from the proximity of trees.

In summary, therefore, the difficulties of applying herbicides and pesticides will certainly reduce the profitability of arable farming, but increases in the populations of insectivorous birds and predatory insects may counteract some of the losses. Silvoarable systems are a natural accompaniment to organic farming, and there is ample opportunity for research on the effects of different agricultural practices on wildlife and on the growth of trees.

Conservation of nutrients

In both silvopastoral and silvoarable systems, the tree component is unlikely to experience a shortage of nutrients. Indeed, it is possible that the high levels of nitrogen applied routinely to reseeded grassland may induce over-rapid growth in some tree species, and lead to a loss of timber strength. Loss of form caused the abandonment of early trials of widely spaced and heavily fertilized trees in Northern Ireland (J H McAdam pers. comm.). The nature and timing of fertilizer applications in agroforests are, therefore, an important topic for research.

Given normal agricultural levels of fertilizer application, and accepting that good agroforest species of tree will root more deeply than the herbaceous intercrops, it appears unlikely that competition for nutrients will be as serious as initial competition for moisture and developing competition for light.

It is possible that some trees may increase the fertility of extensively managed grassland because the roots of most tree species will penetrate to depths which are not accessible to ground vegetation; this is a major criterion in the selection of suitable agroforestry species. Pines, Douglas fir, silver fir (*Abies alba*), oak and sycamore (*Acer pseudoplatanus*) are particularly deep rooting, whereas noble fir (*Abies nobilis*), spruces, western hemlock (*Tsuga heterophylla*) and beech (*Fagus sylvatica*) root near the surface. Deep and widespread tree roots, in both silvoarable and silvopastoral systems, will intercept a proportion of the fertilizer lost to an herbaceous crop. This interception could significantly reduce eutrophication problems in watercourses. More than 50% of the nitrogen applied to pastures in Britain is estimated to be lost through leaching or denitrification, and phosphorus losses from arable areas are becoming increasingly apparent (Frissel 1978).

A significant advantage of energy mixtures, where the understorey is coppiced for fuelwood, may be the interaction between litter of different species. These interactions can speed the decomposition process and increase soil fertility. Birch (*Betula* spp.), for example, tends to develop a mull humus on many soils, with an increased pH, more exchangeable cations and an enhanced earthworm population (Miles 1986). Brown and Dighton (see page 65) have shown the nutritional benefit of growing trees in mixtures, particularly on sites where growth is limited by nitrogen.

Livestock grazing in silvopastoral systems can cause considerable damage to trees by browsing, bark-stripping, or trampling (Adams 1975). However, much of the damage is caused by stocking at excessive rates, planting on soils with impeded drainage which are particularly susceptible to compaction, or using inappropriate species like Sitka spruce which has superficial and easily damaged roots. Animals assist in the recycling of nutrients and often serve to maintain fertility on hill pasture (Floate 1970). However, silvopastoral shelter strips established in larger fields may encourage the excessive congregation of stock, and could suffer damage which would not occur if the whole field were planted with wide-spaced trees.

Hornung and Adamson (see page 57) have shown that clearfelling of forests can cause a significant loss of nutrients in runoff. It also leads to denitrification, ie loss of nitrogen to the atmosphere. Denitrification is favoured by the wet, anaerobic conditions which develop in British uplands following felling, and annual losses of approximately 10 kg N ha^{-1} have recently been measured in Kershope Forest (P Ineson pers. comm.). The felling of agroforests, or the removal of standards in energy coppices, is unlikely to cause any significant loss of nutrients because the roots of the associated species will take up any excess nutrients.

Finally, the prospects for nitrogen-fixing mixtures should be mentioned. Suitable species of alder, or perhaps false acacia (Robinia pseudoacacia) in the south of England, could be grown in row mixtures with crops, and could provide a saving in nitrogen fertilizers. Alder coppice is of proven benefit to intermixed broadleaves such as poplar (de Bell & Radwan 1979). Tree lupin (Lupinus arboreus) has been used as a nitrogen-fixing nurse for conifers and broadleaves (Marrs et al. 1982). Gorse (Ulex europaeus) has been recorded to fix more than 70 kg N ha^{-1} yr^{-1} on mining wastes in Cornwall (Dancer, Handley & Bradshaw 1977). Sea buckthorn (Hippophaë rhamnoides) and broom (Cytisus scoparius) are other nitrogen-fixing possibilities for use in mixture with wide-spaced timber trees.

Soil and water conservation

Of the arable land in England and Wales, 37% is considered susceptible to erosion. To this percentage must be added 7% of upland grazing land. Mean annual rates of soil erosion in fields on hill slopes up to 11 degrees can exceed 2 kg m^{-2}, and rates as high as 1.9 kg m^{-2} have been recorded for individual storms. Where gullying occurs, erosion rates can be much higher, and future arable production is seriously at risk in many parts of the country (Morgan 1985). Trees have a soil conservation role in arable areas which is widely recognized on the Continent. Road, rail and stream margins are often planted with poplars, and 15% of timber production in the Netherlands is gained from such 'linear features' (A Willems pers. comm.). Until recently, line plantings contributed 10% of the total timber production in Italy (Prevosto 1979).

Cumulative shelter is an important attribute of a landscape well-populated by trees, and can be expected as an eventual benefit of widespread agroforestry. Jensen (1954) demonstrated, for example, that the varied topography in central Jutland, with shelterwoods and hedgerows, produces more than twice the reduction in windspeed measured in similar winds blowing across the more featureless landscape in south Jutland.

Soil erosion is associated with the clearfelling of forests and is exacerbated by modern techniques of whole-tree harvesting, which may also cause excessive removal of nutrients and soil acidification (Malkonen 1976). Agroforests will retain a cover of ground vegetation at time of harvest, thereby reducing erosion. Indeed, this is one of the prime reasons for their increasing use in Australia and New Zealand (Reid & Wilson 1985). Tree canopies also limit erosion and the leaching of nutrients by reducing the quantity and intensity of rainfall reaching the ground.

The effect of afforestation on reservoir catchments has been accepted as a problem ever since Law (1956) calculated, in an area receiving 990 mm rainfall annually, that the annual moisture loss from a forest plantation was 290 mm greater than that from grassland. This increased loss is caused by the high interception of rainfall, and subsequent evaporation, from the dense canopies of many conifers. Law subsequently calculated that the water industry lost £500 ha^{-1} yr^{-1} from the afforested parts of a reservoir catchment. A sparse canopy of pruned trees at 10 m spacings will obviously intercept less rainfall than a conventional plantation, and may reduce evaporation from the pasture by providing shelter. It is uncertain, however, whether current models are good enough to predict the scale of this reduction, and experimental studies are needed on catchments with trees planted at agroforestry spacing. The possible hydrological advantage of agroforestry may be an important factor in those catchments which are used for water supply.

Local climate

The effects of a tree canopy on local climate are to:

i. intercept and redirect rainwater, thereby ameliorating excesses and shortages of moisture;

ii. reduce the quantity, and alter the quality, of light reaching the ground;

iii. insulate the field layer and reduce temperature extremes;

iv. reduce windspeed and possible mechanical damage to crops.

These effects on moisture, light and temperature at the ground surface will, in turn, influence the growth of plants and the performance of animals. However, the effect of shelter on the yield of agricultural crops and grasses is still not fully explained, despite many decades of research and speculation. Dramatic increases in crop yield have been reported from windswept continental climates (see reviews of van Eimern *et al.* 1964; Marshall 1967; Sturrock 1984). However, the effects are more moderate in cool oceanic climates. Increases in soil moisture content due to shelter are normally of little importance in the wetter parts of Britain, but higher daytime temperatures in spring can produce a useful 'early bite' of grass growth (Alcock 1969). Seasonal variation makes it difficult to draw conclusions. Alcock, Harvey and Tindsley (1976), working in Wales, found that shelter increased the yield of rye-grass by more than 50% at the end of June, but that there was no effect of shelter at the end of July. In contrast, Russell and Grace (1979) found no effect of shelter on the dry matter production of rye-grass in spring, but recorded a 28% increase during the summer regrowth period. Crop yields in western Britain are likely to benefit less from shelter than in the east. Significant moisture stress commonly develops up to 360–450 m OD in eastern hills (Grant & King 1969), and the shade of deeply rooting trees will be a factor in reducing evaporation.

Several useful attempts have been made to model the effect of trees on crop growth (eg McMurtie & Wolf 1983), but anomalies exist within the literature on shelter effects and highlight the need for practical experimentation in agroforestry. It is even more complicated to predict the multiple consequences of shelter, shade, and competition for moisture and nutrients.

Shelter has a well-established benefit for stock, and this can be crucial during lambing time in the hills and uplands. Although it is possible to establish the metabolic saving to animals due to shelter from the sun (Priestley 1957) or wind (Grace & Easterbee 1979), the possible increases in lambing percentages and live weight gain have not yet been included in any economic models of agroforestry.

14.5 Employment and enjoyment

Whitby (see page 31) has discussed the impact of afforestation on rural employment and has concluded that, where conifer forests replace agriculture, there will be a net loss of jobs over the length of a forest rotation on all but the least intensive of upland agriculture. This conclusion should be contrasted with the analysis of Mutch and Hutchinson (1980), where the establishment of farm forestry on 17 000 ha of hill land caused the numbers of full-time farm employees to decrease from 35.5 to 32, whilst forestry jobs increased from 1.5 to 28. Experience from other countries (eg Prevosto 1976) confirms that agroforestry will largely sustain the agricultural labour force, whilst also creating jobs in forestry and related industries. It is an advantage that farmers and farm workers can use slack time on the farm to manage trees. Agroforestry requires the careful control of stock and cultivation methods. These skills are labour intensive, and will demand enthusiasm for both farming and forestry.

It is harder to predict the consequences for the landscape of widely spaced planting. Neat arrays of pruned clonal conifers could be as oppressive as dense plantations, but agroforestry requires careful management, and it is unlikely to proceed on the scale of plantation forestry. Broadleaved trees can only improve the arable landscape, even in the regular rows used in many parts of Europe.

14.6 Conclusion

The likely advantages and disadvantages of agroforestry are presented in Appendix I. However, many, if not most, of the assumptions in this Appendix require substantiation by further research. It is encouraging, therefore, to report that a co-ordinated programme of research is now under way linking the Agricultural and Food Research Council, the Natural Environment Research Council, the Department of Agriculture for Northern Ireland, and a number of universities.

It is noticeable that temperate agroforestry has been most successful in New Zealand, a country with no state subsidy for agriculture or forestry. The subsidies available in Britain for conventional agriculture or forestry make it difficult for agroforestry to guarantee a comparable financial return.

It will be unfortunate if a grant system in favour of conventional timber species, traditional spacings, and of block sizes exceeding one ha, precludes the options for farmers to integrate arboriculture more closely with cropping or grazing. This presumption also misses an excellent opportunity to encourage the recreation of parklands (Plate 6) and fails to recognize that trees outside the forest, which are properly looked after, can make a valuable multiple contribution to timber production, to the farm enterprise, and to conservation.

Appendix 1.   The possible effects of agroforestry compared with intensive forestry or agriculture

	Yield	Environment	Economics
Advantages	Longer canopy duration Efficient canopy architecture Use of moisture and nutrients at different depths Lessening of climatic extremes Disease less damaging in mixtures Crop fertilizers increase tree yields Wide-spaced trees grow faster Mutual effects of species on nutrient mobilization	More diverse species + structure Less leaching and eutrophication Less wind and water erosion Reduced nutrient loss after felling Less fire risk than forestry Less intensive use of pesticides and herbicides Preservation of rural employment	Several saleable crops More regular revenue than forestry Shared infrastructure costs Less spasmodic use of labour Supplementary sporting income Tree component can be stored to await favourable prices Silvopastoral systems have lower evaporative losses than forestry Wide-spaced trees more windfirm Shelter and shade increase animal production
Disadvantages	Difficult to predict yields in long term Fertility reduced by multiple harvests	Uniform tree rows unattractive Some suggested energy crops are invasive weeds	Silvopastoral fertilizer use may damage timber structure Pesticide applications more difficult Less regular revenue than farming Soil and tree damage by animals Less flexible than farming Dispersed production means high transport costs for wood fuel Greater evaporative losses than farming Greater management effort Cultivation difficult in second rotation

References

Adams, S.N. 1975. Sheep and cattle grazing in forests: a review. *Journal of Applied Ecology*, **12**, 143–152.

Alcock, M.B. 1969. The effect of climate on primary production of temperate grasslands – with reference to upland climate and shelter. *3rd Symposium on Shelter Research*, 49–75. London: Ministry of Agriculture, Fisheries and Food.

Alcock, M.B., Harvey, G. & Tindsley, S.F. 1976. The effect of shelter on pasture production on hill land. *4th Symposium on Shelter Research*, 88–104. London: Ministry of Agriculture, Fisheries and Food.

Anderson, J.A. 1987. Policy options. In: *Agriculture surpluses – environmental implications of changes in farming policy and practice in the UK*, 11–17. London: Institute of Biology.

Arthur-Worsop, M.J. 1985. An economic evaluation of agroforestry. *New Zealand Agricultural Science*, **19**, 99–106.

Booth, T.C. 1988. Agroforestry and growing wood for energy. In: *Farming and forestry*, 95–100. (Forestry Commission occasional paper no. 17.) Edinburgh: Forestry Commission.

Brenchley, A. 1984. The use of birds as indicators of change in agriculture. In: *Agriculture and the environment*, edited by D. Jenkins, 123–127. Cambridge: Institute of Terrestrial Ecology.

Brown, D.A.H. 1988. Land use changes up to the year 2000. In: *Farming and forestry*, 37–39. (Forestry Commission occasional paper no. 17.) Edinburgh: Forestry Commission.

Callaghan, T.V., Lawson, G.J., Millar, A. & Scott, R. 1986. Environmental aspects of agroforestry. In: *Agroforestry – a discussion of research and development required*, 50–76. London: Ministry of Agriculture, Fisheries and Food.

Caborn, J.M. 1965. *Shelterbelts and windbreaks*. London: Faber.

Commission of European Communities. 1986. *Memorandum forestry – discussion paper on the community action in the forest sector*. (Green Europe no. 36.) Luxemburg: EEC Publications Office.

Chaboussou, F. 1986. How pesticides increase pests. *Ecologist*, **16**, 29–35.

de Damas, M. 1978. *Effects des facons culturales sur peuplier 'I-215' en sol semi-terrestre.* Academie d'agriculture de France. Extract du proces verbal de la seance du 8 Novembre 1978, 1224–1229.

Dancer, W.S., Handley, J.F. & Bradshaw, A.D. 1977. Nitrogen accumulation in kaolin mining wastes in Cornwall, England. *Plant and Soil*, **48**, 153–168.

de Bell, D.S. & Radwan, M.A. 1979. Growth and nitrogen relations of coppiced black cottonwood and red alder in pure and mixed plantings. *Botanical Gazette*, **140** (suppl.), 97–101.

Department of Energy. 1987. *Growing wood for energy in Great Britain (the land availability study).* Harwell: Energy Technology Support Unit.

Doyle, C.J., Evans, J. & Rossiter, J. 1987. Agroforestry: an economic appraisal of the benefits of intercropping trees with grassland in lowland Britain. *Agricultural Systems*, **21**, 1–32.

Fauss, D.L. & Pierce, W.R. 1969. Stand conditions and spruce budworm damage in a western Montana forest. *Journal of Forestry*, **67**, 322–325.

Fenton, R. & Sutton, W.R.J. 1968. Silvicultural proposals for radiata pine on high quality sites. *New Zealand Journal of Forestry*, **13**, 220–228.

Filius, A.M. 1982. Economic aspects of agroforestry. *Agroforestry Systems*, **1**, 29–39.

Floate, M.J.S. 1970. Mineralisation of nitrogen and phosphorus from organic materials of plant and animal origin and its significance in the nutrient cycle in grazed upland and hill soils. *Journal of the British Grassland Society*, **25**, 295–302.

Food and Agriculture Organization. 1980. *Poplars and willows in wood production and land use.* Rome: FAO.

Frissel, M.J. 1978. *Cycling of mineral nutrients in agricultural ecosystems.* Amsterdam: Elsevier.

Grace, J. & Easterbee, N. 1979. The natural shelter for red deer (*Cervus elaphus*) in a Scottish glen. *Journal of Applied Ecology*, **16**, 37–48.

Grant, S. & King, J. 1969. Altitude and grass growth – some observations on apparent exposure effects. *3rd Symposium on Shelter Research*, 77–86. London: Ministry of Agriculture, Fisheries and Food.

Hardy, A.R. 1986. The Boxworth project – a progress report. In: *1986 British Crop Protection Conference*, 1215–1224. London: British Crop Protection Council.

Jensen, M. 1954. *Shelter effect.* Copenhagen: Danish Technical Press.

Law, F. 1956. The effect of afforestation upon the yield of water catchment areas. *Journal British Waterworks Association*, 489–494.

Lawson, G.J. 1987. Hill weed compensatory allowances: very alternative crops for the uplands. In: *Agriculture and conservation in the hills and uplands*, edited by M. Bell & R.G.H. Bunce, 99–106. (ITE symposium no. 23.) Grange-over-Sands: Institute of Terrestrial Ecology.

Malkonen, E. 1976. The effect of fuller biomass harvesting on soil fertility. In: *Symposium on the harvesting of a larger part of the forest biomass.* Hyvinkaa, Finland: Economica Commission for Europe/Food and Agriculture Association/International Labour Association.

Marrs, R.H., Owen, L.D.C., Roberts, R.D. & Bradshaw, A.D. 1982. Tree lupin (*Lupinus arboreus* Sims): an ideal nurse crop for land restoration and amenity plantings. *Arboricultural Journal*, **6**, 161–174.

Marshall, J.K. 1967. The effect of shelter on the productivity of grasslands and field crops. *Field Crop Abstracts*, **20**, 1–14.

Maxwell, T.J. 1986. Agroforestry systems for hills and uplands. In: *Agroforestry – a discussion of research and development requirements.* London: Ministry of Agriculture, Fisheries and Food.

McElroy, G.H. & Dawson, W.M. 1986. *Production and utilisation of biomass from short rotation coppices willow in northern Ireland 1975–1985.* Loughall: Department of Agriculture for Northern Ireland, Horticultural Centre.

McMurtie, R.S. & Wolf, L. 1983. A model of competition between trees and grass for radiation, water and nutrients. *Annals of Botany*, **52**, 449–458.

Miles, J. 1986. What are the effects of trees on soils? In: *Trees and wildlife in the Scottish uplands*, edited by D. Jenkins, 55–62. (ITE symposium no. 17.) Abbots Ripton: Institute of Terrestrial Ecology.

Ministry of Agriculture, Fisheries and Food. 1986. *Annual review of agriculture.* London: HMSO.

Mitchell, C.P. 1987. Investigations on short rotation forestry for energy. In: *Energy from biomass*, edited by G. Grassi & H. Zibetta, 36–40. London: Applied Science.

Morgan, R.P.C. 1985. Soil erosion in Britain: the loss of a resource. *Ecologist*, **16**, 40–41.

Mutch, W.E.S. & Hutchinson, A.R. 1980. *The interaction of forestry and farming.* (Economics & Management Series 2.) Edinburgh: East of Scotland College of Agriculture.

Newton, I. & Moss, D. 1981. Factors affecting the breeding of sparrowhawks and the occurrence of their songbird prey in woodlands. In: *Forest and woodland ecology*, edited by F.T. Last & A.S. Gardiner, 125–131. (ITE symposium no. 8.) Cambridge: Institute of Terrestrial Ecology.

Niemeyer, H. 1986. Managing forests for wildlife in Germany. In: *Trees and wildlife in the Scottish uplands*, edited by D. Jenkins, 158–165. (ITE symposium no. 17.) Abbots Ripton: Institute of Terrestrial Ecology.

O'Carroll, N. 1982. The nursing of Sitka spruce 2. Nitrogen-fixing species. *Irish Forestry*, **39**, 17–29.

Penalosa, R., Herve, M. & Sobarzo, L. 1985. Applied research on multiple land use through silvopastoral systems in southern Chile. *Agroforestry Systems*, **3**, 59–77.

Percival, N.S. & Hawke, M.F. 1985. Agroforestry development and research in New Zealand. *New Zealand Agricultural Science*, **19**, 86–92.

Prevosto, M. 1976. The economics of extensive poplar culture in Italy. *Proceedings Symposium on eastern cottonwood and related species*, 438–445. Baton Rouge: Louisiana State University.

Prevosto, M. 1979. Growth and revenues of poplars grown in specialized stands subjected or not to thinning. In: *FAO–IUFRO technical consultation on fast-growing plantation broadleaved trees for Mediterranean and temperate zones.* Lisbon: Food and Agriculture Organisation.

Price, R. & Mitchell, C.P. 1985. *Potential for wood as fuel in the United Kingdom.* (ETSU R32.) London: Department of Energy.

Priestley, C.H.B. 1957. The heat balance of a sheep standing in the sun. *Australian Journal of Agricultural Research*, **8**, 271.

Reed, J. 1982. Birds and afforestation. *Ecos*, **3**, 8–10.

Reid, R. & Wilson, G. 1985. *Agroforestry in Australia and New Zealand*. Victoria: Goddard & Dobson.

Russell, R.C. 1947. The chemical composition and digestibility of fodder shrubs and trees. *Joint Publications, Imperial Agricultural Bureaux*, **10**, 185–231.

Russell, G. & Grace, J. 1979. The effect of shelter on the yield of grasses in southern Scotland. *Journal of Applied Ecology*, **16**, 319–330.

Savill, P.S. & Evans, J. 1986. *Plantation silviculture in temperate regions*. Oxford: Clarendon Press.

Scott, R., Watt, C.P., Buckland, M.P., Lawson, G.J. & Callaghan, T.V. 1986. Biofuels: the domestic market in the UK. In: *Energy for rural and island communities, IV*, edited by J. Twidell, I. Hounam & C. Lewis, 115–120. Oxford: Pergamon.

Sibbald, A.R., Maxwell, T.J., Griffiths, J.H., Hutchings, N.J., Taylor, C.M.A., Tabbush, P.M. & White, I.M.S. 1987. Agroforestry research in the hills and uplands. In: *Agriculture and conservation in the hills and uplands*, edited by M. Bell & R.G.H. Bunce, 74–77. (ITE symposium no. 23.) Grange-over-Sands: Institute of Terrestrial Ecology.

Staines, B.W. 1986. Mammals of Scottish upland woods. In: *Trees and wildlife in the Scottish uplands*, edited by D. Jenkins, 112–120. (ITE symposium no. 17.) Abbots Ripton: Institute of Terrestrial Ecology.

Sturrock, J.W., ed. 1984. *Shelter research needs in relation to primary production. Report of the National Shelter Working Party*. Wellington: Water and Soil Conservation Authority.

van Eimern, J., Karschow, R., Razumova, L.A. & Robertson, G.W. 1964. *Windbreaks and shelterbelts*. (Technical notes no. 59.) Geneva: World Meteorological Organisation.

Zsuffa, L. & Barkley, B. 1985. The commercial and practical aspects of short-rotation forestry in temperate regions: a state-of-the-art review. In: *Bioenergy '84. Vol. 1*, edited by H. Egneus & A. Elkgard, 39–57. London: Elsevier.

Appendix: List of participants

J K Adamson, Institute of Terrestrial Ecology, Merlewood Research Station, Grange-over-Sands, Cumbria.

S E Allen, Institute of Terrestrial Ecology, Merlewood Research Station, Grange-over-Sands, Cumbria.

A R Anderson, Forestry Commission, Northern Research Station, Roslin, Midlothian.

D Andrews, British Trust for Conservation Volunteers, National Park Centre, Brockhole, Windermere, Cumbria.

N Banister, Timber Growers UK Ltd, Ympgarth House, Stainton, Penrith, Cumbria.

A Barker, British Trust for Conservation Volunteers, National Park Centre, Brockhole, Windermere, Cumbria.

J M Barratt, 2 Scarfoot Cottages, Meal Bank, Kendal, Cumbria.

J Beckett, Institute of Terrestrial Ecology, Merlewood Research Station, Grange-over-Sands, Cumbria.

R Bell, Robin Forestry Surveys, 26 Harker Park Road, Carlisle, Cumbria.

S Bell, Country Landowners Association, 16 Belgrave Square, London SW1.

D Birkett, Lake District Ranger Service, Oakmount, 206 Burneside Road, Kendal, Cumbria.

G Booth, Fisher Hoggarth, Land Agents, The Old Rectory, Hutton-in-the-Forest, Penrith, Cumbria.

D R Briggs, Institute of Terrestrial Ecology, Merlewood Research Station, Grange-over-Sands, Cumbria.

A H F Brown, Institute of Terrestrial Ecology, Merlewood Research Station, Grange-over-Sands, Cumbria.

J K Buchanan, Countryside Commission, 4th Floor, Warwick House, Grantham Road, Newcastle-upon-Tyne.

M Buckton, Cheviot Forestry Consultants Ltd, Greystoke Forest Office, Hutton Roof, Penrith, Cumbria.

R G H Bunce, Institute of Terrestrial Ecology, Merlewood Research Station, Grange-over-Sands, Cumbria.

T V Callaghan, Institute of Terrestrial Ecology, Merlewood Research Station, Grange-over-Sands, Cumbria.

R N Cartwright, County Planning Department, Cumbria County Council, County Offices, Kendal, Cumbria.

E Clark, Old Brathay, Ambleside, Cumbria.

J Clark, National Trust, Rothay Holme, Rothay Road, Ambleside, Cumbria.

R Clark, Parcey House, Hartsop, Penrith, Cumbria.

G E Curzon, 11 Grizedale Close, Keswick, Cumbria.

E W Curzon, 11 Grizedale Close, Keswick, Cumbria.

D E Davis, Forestry Department, Cumbria College of Agriculture & Forestry, Newton Rigg, Penrith, Cumbria.

P A Dawson, Ministry of Agriculture, Fisheries and Food, Eden Bridge House, Lowther Street, Carlisle, Cumbria.

J Dighton, Institute of Terrestrial Ecology, Merlewood Research Station, Grange-over-Sands, Cumbria.

J Dodgson, Country Landowners Association, Gurnel Beck, Millside, Witherslack, Grange-over-Sands, Cumbria.

I M Dowsing, Flat 5, Sunny Brae, Rockland Road, Grange-over-Sands, Cumbria.

R D Everett, Middle Wood Centre, Roeburndale West, Wray, Lancaster, Lancashire.

T Farrow, County Planning Department, Cumbria County Council, County Offices, Kendal, Cumbria.

A Fishwick, Lake District National Park Authority, Busher Walk, Kendal, Cumbria.

J A Galloway, 36 Victoria Road North, Windermere, Cumbria.

L Gee, Flat 1, East Bank, Eden Mount Road, Grange-over-Sands, Cumbria.

M H Gee, 19 Castle Road, Kendal, Cumbria.

T Goddard, Middle Wood Centre, Roeburndale West, Wray, Lancaster, Lancashire.

J Gurney, National Trust, Rothay Holme, Rothay Road, Ambleside, Cumbria.

D Hickson, 18 Carlisle Road, Dalston, Carlisle, Cumbria.

T Hill, The Woodland Trust, 1 West Lodge, Sutton Lane, Babworth, Retford, Nottinghamshire.

J R Hooson, Nature Conservancy Council, Blackwell, Bowness-on-Windermere, Cumbria.

M Hornung, Institute of Terrestrial Ecology, Penrhos Road, Bangor, Gwynedd.

J M Houston, Friends of the Lake District, Gowan Knott, Kendal Road, Staveley, Kendal, Cumbria.

P Ineson, Institute of Terrestrial Ecology, Merlewood Research Station, Grange-over-Sands, Cumbria.

A J Kerby, 4 Cavendish Street, Ulverston, Cumbria.

D Kerr, Forestry Commission Research Branch, Kielder, Hexham, Northumberland.

K J Kirby, Nature Conservancy Council, Northminster House, Northminster, Peterborough, Cambridgeshire.

P M Latter, Institute of Terrestrial Ecology, Merlewood Research Station, Grange-over-Sands, Cumbria.

G J Lawson, Institute of Terrestrial Ecology, Merlewood Research Station, Grange-over-Sands, Cumbria.

P J W S Ling, Watson Lewis & Co, St Andrew's Churchyard, Penrith, Cumbria.

G B Little, Thames Harvesting, Workington, Cumbria.

A J Low, Forestry Commission, Northern Research Station, Roslin, Midlothian.

R McGuffie, Forestry Office, North West Water, Thirlmere, Keswick, Cumbria.

P M McMurdo, YMCA, Lakeside, Ulverston, Cumbria.

A S C Meikle, Economic Forestry plc, 95 Highgate, Kendal, Cumbria.

R J Metcalfe, YMCA, Lakeside, Ulverston, Cumbria.

M Mills, Low Gillerthwaite Field Centre, Ennerdale, Cleator, Cumbria.

W Milne, Museum of Natural History and Archaeology, Station Road, Kendal, Cumbria.

F Mitchell, Hill Farming Research Organisation, Bush Estate, Penicuik, Midlothian.

R Mitchell, Cat Nest, Grizebeck, Kirkby-in-Furness, Cumbria.

D Moss, County Planning Department, Cumbria County Council, County Offices, Kendal, Cumbria.

K Parker, National Trust, Rothay Holme, Rothay Road, Ambleside, Cumbria.

S Preston, Yorkshire Dales National Park, Colvend, Hebden Road, Grassington, North Yorkshire.

S L Reynolds, Department of Geography, University of Lancaster, Lancaster, Lancashire.

T Riden, 3 Makinson Row, Galgate, Lancaster, Lancashire.

J E Satchell, Draw Well, Lyth, Kendal, Cumbria.

A C Seale, 5 St Anne's Close, Ambleside, Cumbria.

R E Shapland, Forestry Office, North West Water, Thirlmere, Keswick, Cumbria.

F J Shaw, Institute of Terrestrial Ecology, Merlewood Research Station, Grange-over-Sands, Cumbria.

M Sim, National Trust, Rothay Holme, Rothay Road, Ambleside, Cumbria.

G F Smart, Holker Woodlands, Holker, Cark-in-Cartmel, Grange-over-Sands, Cumbria.

W Smith, Flat 5, Sunny Brae, Rockland Road, Grange-over-Sands, Cumbria.

J M Sykes, Institute of Terrestrial Ecology, Merlewood Research Station, Grange-over-Sands, Cumbria.

D Thomason, National Trust, Rothay Holme, Rothay Road, Ambleside, Cumbria.

J C Voysey, Forestry Commission, Grizedale, Hawkshead, Ambleside, Cumbria.

J Walne, Ministry of Agriculture of Fisheries and Food, Eden Bridge House, Lowther Street, Carlisle, Cumbria.

A M Whitbread, Nature Conservancy Council, 70 Castlegate, Grantham, Lincolnshire.

M Whitby, Department of Agricultural Economics, University of Newcastle-upon-Tyne, Newcastle-upon-Tyne.

R Williamson, St Oswald's Vicarage, Burneside, Kendal, Cumbria.

P L Winchester, Royal Forestry Society, Whitefoot Cottage, Burneside, Kendal, Cumbria.

R Wormald, Youth Hostels Association, 78 Oxenholme Road, Kendal, Cumbria.

J D Young, Dunans, Cotehill, Carlisle, Cumbria.

Printed in the United Kingdom for Her Majesty's Stationery Office
Dd 291030   C20   8/89   59471